FLORA OF THE ~~~ ^ C

Edited b

M.J. JANSEN-JACOBS

Series A: Phanerogams

Fascicle 26

155. GESNERIACEAE
(L.E. Skog & C. Feuillet)

2008
Royal Botanic Gardens, Kew

Contents

155. GESNERIACEAE

by

LAURENCE E. SKOG AND CHRISTIAN FEUILLET[1]

Terrestrial or epiphytic perennial herbs, subshrubs, shrubs, small trees or lianas, sometimes arising from tubers, scaly rhizomes or stolons, rarely acaulescent; above ground stems sometimes annual, usually terete, erect, ascending, pendent or scandent, branched or not, usually pubescent at least towards apex, and often producing adventitious roots. Leaves estipulate, usually petiolate, opposite, rarely whorled or alternate, simple; blades equal to strongly unequal in a pair, membranous, fleshy or coriaceous, entire to variously toothed, often at least pubescent, stomata scattered on lower leaf surface or rarely in clusters (only *Napeanthus*). Inflorescences derived from cymes, but sometimes in modified racemes, occasionally fasciculate, or reduced to a single flower, axillary or terminal, usually pedunculate, bracteate or not. Flowers perfect, often protandrous, usually zygomorphic, rarely nearly actinomorphic; calyx lobes free or connate, (2 or 4)5, equal or unequal, usually imbricate, green or colored, entire or variously toothed; corolla with a short or long tube of 5 connate petals, variously colored, often oblique in calyx and gibbous to saccate at base, cylindric, ventricose, or ampliate above, limb of (4)5 equal or unequal lobes, sometimes 2-lipped, erect or spreading to reflexed; stamens usually adnate to base of corolla tube, 4, with or without staminode and usually didynamous, or rarely 5, included or exserted, usually coiling or retracting after pollen shedding, anthers more or less coherent at first, thecae dehiscing by pores or longitudinal slits; disc annular, or of 1-5 separate or connate nectariferous glands, or rarely absent; pistil 2-carpellate, ovary superior to inferior, 1-loculed, placentae parietal, ovules numerous, style simple, stigma usually 2-lobed or stomatomorphic. Fruit a berry, or a dry or fleshy capsule, 2- or 4-valved; seeds numerous, more or less fusiform or oblong, very small, usually striate.

[1] Department of Botany, MRC-166, Smithsonian Institution, P.O. Box 37012, Washington, DC 20013-70162, U.S.A.

We gratefully acknowledge: the artists for the preparation of the illustrations Cathy Pasquale-Johnson, Ellen Seefelt, and Alice Tangerini whose names are mentioned on the figures they drew, and Amy Melson (AM); the encouragement and help by Anton J.M. Leeuwenberg, Wageningen University, The Netherlands; the assistance of John K. Boggan in the preparation of the manuscript; the Missouri Botanical Garden and the New York Botanical Garden for permission to use illustrations from Novon and Brittonia, respectively; and the financial support of the Biodiversity of the Guianas Program and other funding sources at the U.S. National Museum of Natural History.

Distribution: Widely distributed in tropical regions of the world, rarely reaching into warm temperate areas; about 2600 species in about 130 genera; in the Neotropics about 1200 species; in the Guianas 60 species in 20 genera.

Notes: Plants of GESNERIACEAE are widely cultivated as pot-plants or ornamentals, and known mostly for the 'African Violet' (*Saintpaulia spp.* and cultivars) and the 'Florist's Gloxinia' (*Sinningia speciosa* (Lodd.) Hiern and cultivars). Occasionally escaping from cultivation in the tropics, plants rarely become naturalized in the Guianas.

LITERATURE

Aublet, J.B.C.F. 1775. Histoire des plantes de la Guiane Françoise, 4 vols.

Feuillet C. & L.E. Skog. 2002. Gesneriaceae. In S.A. Mori *et al.*, Guide to the Vascular Plants of Central French Guiana, 2. Mem. New York Bot. Gard. 76(2): 334-344.

Feuillet C. & L.E. Skog. 2003a ('2002'). Novae Gesneriaceae Neotropicarum, 11. Brittonia 54: 344-351.

Feuillet C. & L.E. Skog. 2003b ('2002'). Novae Gesneriaceae Neotropicarum, 12. Brittonia 54: 352-361.

Feuillet, C. & J.A. Steyermark. 1999. Gesneriaceae. In J.A. Steyermark *et al.*. Flora of the Venezuelan Guayana 5: 542-573.

Fritsch, K. 1893-1894. Gesneriaceae. In H.G.A. Engler & K.A.E. Prantl, Die natürlichen Pflanzenfamilien 4(3b): 133-144. 1893; 145-185. 1894.

Gibson, D.N. 1972. Studies in American plants, 4. Phytologia 23: 334-342.

Gibson, D.N. 1974. Gesneriaceae. In P.C. Standley & L.O. Williams, Flora of Guatemala, 10(3). Fieldiana, Bot. 24(10): 240-313.

Graham, E.H. 1933. Gesneriaceae. In Flora of the Kartabo region, British Guiana. Ann. Carnegie Mus. 22: 250-251.

Grenand, P. *et al.* 1987. Pharmacopées traditionnelles en Guyane, Gesneriaceae. pp. 240-242.

Grenand, P. *et al.* 2004. Pharmacopées traditionnelles en Guyane, ed. 2, Gesneriaceae. pp. 379-382.

Hanstein, J. 1854-1865. Die Gesneraceen des Königlichen Herbariums und der Gärten zu Berlin. Linnaea 26: 145-216. 1854 ('1853'); 27: 693-785. 1856 ('1854'); 29: 497-592. 1859 ('1857-1858'); 34: 225-462. 1865 ('1864-1865').

Hanstein, J. 1864. Gesneriaceae. In C.F.P. von Martius, Flora Brasiliensis 8(1): 343-428.

Kvist, L.P. 1986. Gesneriads and snake bite. Gloxinian 36(1): 8-13.

Leeuwenberg, A.J.M. 1958. The Gesneriaceae of Guiana. Acta Bot. Neerl. 7: 291-444.

Leeuwenberg, A.J.M. 1984. Gesneriaceae. In A.L. Stoffers & J.C. Lindeman, Flora of Suriname 5(1): 592-650.

Morton, C.V. 1938. Gesneriaceae. In P.C. Standley, Flora of Costa Rica. Publ. Field Mus. Nat. Hist., Bot. Ser. 18: 1137-1187.

Morton, C.V. 1944. Taxonomic studies of tropical American plants. Contr. U.S. Natl. Herb. 29: 1-86.

Morton, C.V. 1948. Gesneriaceae. In B. Maguire *et al.*, Plant Explorations in Guiana in 1944, chiefly to the Tafelberg and the Kaieteur Plateau, 5. Bull. Torrey Bot. Club 75: 563-566.

Morton, C.V. 1953. Gesneriaceae. In J.A. Steyermark *et al.*, Contributions to the flora of Venezuela. Fieldiana, Bot. 28: 520-534.

Morton, C.V. 1954. Gesneriaceae. In Flora of Trinidad and Tobago 2: 301-315.

Morton, C.V. 1957. Gesneriaceae. In H. León & H. Alain, Flora de Cuba 4: 451-472.

Morton, C.V. 1962. Gesneriaceae. In J.A. Steyermark & S. Nilsson, Botanical novelties in the region of Sierra de Lema, Estado Bolivar, 1. Bol. Soc. Venez. Cienc. Nat. 23: 76-83.

Schomburgk, Ri. 1849 ('1848'). Reisen in Britisch-Guiana, Gesneriaceae 3: 971-972.

Skog, L.E. 1979 ('1978'). Gesneriaceae. In R.E. Woodson *et al.*, Flora of Panama, 9. Ann. Missouri Bot. Gard. 65: 783-996.

Skog, L.E. 1984. A Review of chromosome numbers in the Gesneriaceae. Selbyana 7: 252273.

Skog, L.E. 2001. Gesneriaceae. In: W.D. Stevens *et al.*, Flora de Nicaragua. Monogr. Syst. Bot. Missouri Bot. Gard. 85: 1114-1128.

Wiehler, H. 1973. One hundred transfers from Alloplectus and Columnea (Gesneriaceae). Phytologia 27: 309-329.

Wiehler, H. 1978. The genera Episcia, Alsobia, Nautilocalyx, and Paradrymonia (Gesneriaceae). Selbyana 5: 11-60.

Wiehler, H. 1983. A synopsis of the neotropical Gesneriaceae. Selbyana 6: 1-219.

Wiehler, H. 1995. Medicinal gesneriads. Gesneriana 1: 98-120.

Wilson, C. 1974. Floral anatomy in Gesneriaceae, 2. Gesnerioideae. Bot. Gaz. 135: 256-268.

KEY TO THE GENERA

1 Lateral aerial branches with basal internodes much elongated in form of stolons .... *10. Episcia*
  Lateral branches not elongated at their bases into stolons 2

2 Plants stemless; inflorescences and leaves appearing to arise directly from tuber 3
  Plants with obvious stems; inflorescences axillary or terminal 4

3 Leaf venation subpalmate .... *14. Lembocarpus*
  Leaf venation pinnate .... *18. Rhoogeton*

4 Stems much shorter than leaves or with leaves crowded at apex 5
  Stems becoming longer than leaves and leaves not crowded at apex ... 10

5 Corollas red .... *16. Nautilocalyx p.p.*
  Corollas white, sometimes with lobes lavender 6

6 Peduncles about as long or longer than leaves and pedicels very short ....
  ....*13. Lampadaria*
  Peduncles shorter than leaves and pedicels obviously present 7

7 Leaf blades round to acute at base 8
  Leaf blades long attenuate at base 9

8 Leaf blades with indumentum .... *16. Nautilocalyx p.p.*
  Leaf blade glabrous or nearly so .... *17-5. Paradrymonia densa*

9 Leaves sessile or subsessile; inflorescences bracteate .... *15. Napeanthus*
  Leaves obviously petiolate; inflorescences ebracteate ..... *20. Tylopsacas*

10 Plants terrestrial, decumbent to erect 11
  Plants epiphytic or if terrestrial then lianescent or clambering 20

11 Ovary almost completely inferior; plants rhizomatous 12
  Ovary superior or almost completely superior (*Sinningia*); plants with tubers, or lacking underground perennial stems, not rhizomatous 13

12 Corollas purple or if orange-red then lobes tiny and green ... *11. Gloxinia*
  Corollas orange, including spreading lobes .... *12. Kohleria*

13 Inflorescences lacking bracts and always with peduncles (except *Besleria insolita*) and pedicels .... *2. Besleria*
  Inflorescences with bracts, or lacking peduncles 14

14 Stems erect at apex, decumbent at base 15
  Stems erect 16

15 Leaf blades with indumentum .... *16. Nautilocalyx p.p.*
  Leaf blade glabrous or nearly so .... *17-5. Paradrymonia densa*

16 Inflorescence clearly pedunculate 17
   Inflorescence epedunculate or obscurely pedunculate 19

17 Corollas spurred at base, strongly oblique in calyx .... *8. Cremersia*
   Corollas not spurred at base, erect in calyx 18

18 Plants from tubers; fruit a capsule .... *3. Chrysothemis*
   Plants without tubers; fruit a berry .... *7. Corytoplectus*

19 Herbs or shrubs without tubers; corolla yellow; fruit a red berry ....
   .... *2-2. Besleria insolita*
   Herbs from tubers; corollas orange to red; fruit a dry green to brown capsule
   .... *19. Sinningia*

20 Anthers dehiscing by pores; stems glabrous, except sometimes at apex ... 21
   Anthers dehiscing by longitudinal slits; stems usually pubescent 22

21 Fruit a red to dark red berry; extrafloral nectaries present ... *4. Codonanthe*
   Fruit a capsule and often brightly colored inside; no extrafloral nectaries ....
   .... *9. Drymonia*

22 Corollas white or whitish, rarely yellow or cream-colored (in *Paradrymonia*
   maculata and then 4 lobes spreading and the 5th covering mouth of corolla
   tube, and nearly equal leaves) 23
   Corollas red or yellow 25

23 Leaves equal to unequal in a pair, smaller one shaped like larger one, not
   deciduous .... *17. Paradrymonia p.p.*
   Leaves strongly unequal in a pair, smaller one usually stipule-like or
   deciduous 24

24 Ventral lobe of corolla entire .... *5. Codonanthopsis*
   Ventral lobe of corolla long fimbriate .... *17. Paradrymonia p.p.*

25 Calyx lobes ovate to cordate at base, if corolla red then lower lobe yellow
   and not reflexed; fruit a green fleshy capsule .... *1. Alloplectus*
   Calyx lobes narrowed at base, or if cordate then all 5 corolla lobes red and
   lower lobe reflexed (*Columnea oerstediana*); fruit a white or colored
   berry .... *6. Columnea*

1. **ALLOPLECTUS** Mart., Nov. Gen. Sp. Pl. 3: 53. 1829, nom. cons.
   Type: A. hispidus (Kunth) Mart. (Besleria hispida Kunth), typ. cons.

   Crantzia Scop., Intr. Hist. Nat. 173. 1777.
   Type: C. cristata (L.) Fritsch (Besleria cristata L.) [Alloplectus cristatus
   (L.) Mart.]

Terrestrial or epiphytic, caulescent, clambering or erect, coarse herbs or shrubs, without modified stems. Stems rarely branched. Leaves opposite, usually equal in a pair, venation pinnate, foliar nectaries absent. Flowers solitary or in fasciculate few-flowered, epedunculate inflorescences; bracteoles small; pedicellate. Calyx lobes 5, nearly free, ovate to cordate at base; corolla yellow or red, tubular, limb of 5 lobes; stamens included, filaments basally connate, anthers coherent or becoming free, dehiscing by longitudinal slits, thecae parallel; staminode small; disc a single dorsal, large, sometimes 2-lobed gland; ovary superior, stigma stomatomorphic to 2-lobed. Fruit a fleshy, green capsule, loculicidally dehiscent, 2-valved, valves opening slightly.
Chromosome number n=9 (Skog 1984).

Distribution: A genus of about 60 species mainly in C America and northern S America, from Mexico to Bolivia and Suriname, as well as Hispaniola; mostly growing in lowland to mid-elevation rain or cloud forest, and occasionally at forest margins; 2 species in the Guianas.

Use: Plants of *Alloplectus* are rarely cultivated and have no known other economic uses.

Note: *Alloplectus meridensis* Klotzsch ex Hanst. was included by Leeuwenberg (1958: 363) based on a sterile specimen from the Wilhelmina Mts. in Suriname. This specimen was later identified as *A. savannarum* C.V. Morton. *Alloplectus meridensis* is not otherwise expected to be found in the Guianas and is not included here.

LITERATURE

Clark, J.L. 2005. A Monograph of Alloplectus (Gesneriaceae). Selbyana 25: 182-209.

Feuillet, C. & L.E. Skog. 1990. Proposal to emend the type citation of 7860 Alloplectus Mart., nom. cons. (Gesneriaceae: Gesnerioideae). Taxon 39: 133-134.

Morley, B.D. 1974. A Revision of Caribbean species in the genera Columnea L. and Alloplectus Mart. (Gesneriaceae). Proc. Roy. Irish Acad., B 74(24): 411-438.

Stearn, W.T. 1969. The Jamaican species of Columnea and Alloplectus. Bull. Brit. Mus. (Nat. Hist.), Bot. 4: 181-236.

KEY TO THE TAXA

1 Pedicel long, often almost equaling length of leaves; leaves of a pair equal or subequal; corolla nearly 2 times as long as calyx ....
.... *1. A. cristatus* var. *epirotes*
Pedicel shorter than petioles; leaves of a pair unequal, smaller one in a pair ca. $^1/_2$ the size of the larger; corolla only as long as calyx ....
.... *2. A. savannarum*

1. **Alloplectus cristatus** (L.) Mart., Nov. Gen. Sp. Pl. 3: 189. 1832. – *Besleria cristata* L., Sp. Pl. 619. 1753. – *Columnea cristata* (L.) Kuntze, Revis. Gen. Pl. 2: 471. 1891. – *Crantzia cristata* (L.) Fritsch in Engl. & Prantl, Nat. Pflanzenfam. 4(3b): 168. 1894. Type: [icon] Plumier, Pl. Amer. pl. 50. 1756 (neotype, designated by Leeuwenberg 1958: 361).

In the Guianas only: var. **epirotes** Leeuwenb., Acta Bot. Neerl. 7: 298, 363. 1958. Type: Guyana, Cuyuni-Mazaruni Region, Pakaraima Mts., Mt. Ayanganna, Maguire *et al.* 40594 (holotype NY, isotypes NY, U, US).

Epiphytic subshrub, woody climber or liana, to 3 m. Stem woody at base, scandent, glabrescent below, hirsutulous at apex. Leaves equal or subequal in a pair; petiole 0.3-3.5 cm long, densely strigose to subtomentose, densely pilosulous; blade papyraceous when dry, oblong-elliptic to oblong-ovate, 3-10 x 1.5-4.5 cm, margin crenate-serrulate, apex acute to acuminate, base somewhat rounded, above strigose or substrigose, below strigillose, hirsutulous on midvein and veins. Flowers solitary in reduced cymes; pedicel red, 2.5-8 cm long, hirsutulous. Calyx angulate, pale pink to dark red, lobes connate at base, tube ca. 0.3 cm long, free portion of lobes erect, 4 subequal and dorsal lobe smaller and narrower, broadly ovate, 1.2-2 x 2 cm, margin sharply dentate-serrate, apex acute to acuminate, on both sides strigillose; corolla oblique in calyx, yellow, 2.5-3.5 cm long, tube cylindric, 2-3 cm long, base gibbous posteriorly, 0.3-0.5 cm wide, middle subventricose, throat slightly constricted, 0.5-0.8 cm wide, outside densely pilose, pilosulous, inside glabrous, limb 0.9-1.4 cm wide, lobes subequal, spreading, rounded, 0.2-0.4 x 0.25-0.4 cm wide, margin minutely dentate; stamens adnate ca. 0.4 cm to base of corolla tube; ovary ovoid, 0.4-0.5 x 0.3 cm, strigillose, style ca. 2.5 cm long, glabrous, stigma stomatomorphic. Mature capsule white, globose, ca. 1 x 1 cm.

Distribution: Apparently endemic to western Guyana; in wet montane forests at 700-1650 m alt.; 5 specimens examined (GU: 5).

Specimens studied: Guyana: Cuyuni-Mazaruni Region: Pakaraima Mts., Mt. Ayanganna, Maguire *et al.* 40594 (NY, U, US); Potaro-Siparuni Region: Pakaraima Mts., Mt. Wokomung, Boom & Samuels 9197 (US), Henkel *et al.* 1344 (BRG, MO, NY, US), 4435 (CAY, F, NY, US); Mt. Kopinang, Hahn *et al.* 4281 (US).

Notes: The species has 2 varieties, only one of which occurs in the Guianas. The typical variety of *Alloplectus cristatus* is known from the Lesser Antilles. A Leblond collection (at P-LA) of *A. cristatus* var. *cristatus* is labeled as coming from French Guiana. However, Leeuwenberg (1958: 363) considered this locality doubtful as the typical variety is known otherwise only from Martinique and St. Lucia. It differs from var. *epirotes* in having laciniate and serrate calyx lobes.
Leeuwenberg (1958: 363, 368) also noted that the specimen that Aublet (Hist. Pl. Guiane, 1775: 637) identified as *Besleria cristata* is *B. flavovirens* Nees & Mart.
Morley (1974: 419) included var. *epirotes* in *Alloplectus cristatus* (L.) Mart. var. *brevicalyx* C.V. Morton.
Clark (2005) revived *Crantzia*, including among others *C. cristata*, with var. *epirotes* raised to specific rank as *C. epirotes* (Leeuwenb.) J.L. Clark.

2. **Alloplectus savannarum** C.V. Morton in Maguire *et al.*, Bull. Torrey Bot. Club 75: 563. 1948. Type: Guyana, Potaro-Siparuni Region, Kaieteur savannahs, Maguire & Fanshawe 23127 (holotype NY, isotypes A, BR, F, G, K, MO, P, U, UC, US, VEN, W). – Fig. 1

Columnea steyermarkii C.V. Morton in Steyerm. *et al.*, Bol. Soc. Venez. Ci. Nat. 23: 76. 1962. Type: Venezuela, Steyermark & Nilsson 41 (holotype US, isotypes NY, VEN), syn. nov.

Epiphytic liana or subshrub, to 5 m tall. Stem woody at base, scandent, yellow-tomentose or reddish brown with white hairs near apex, glabrescent below. Leaves mostly unequal in a pair; petiole 2-5.5 cm long, yellow-tomentose; blade papyraceous when dry, obliquely oblong-elliptic to broadly elliptic, larger blade 7.5-16 x 3-7 cm, margin serrate, apex shortly acuminate, base obliquely cuneate, not decurrent, above pilosulous, below hirsute to pilose. Flowers 1-2 in reduced cymes; pedicel 0.9-1.2 cm long, hirsute to densely tomentose. Calyx angulate, red, lobes free, erect, subequal, ovate, 2-3 x 1-1.25(-2) cm, margin dentate-serrate to entire at apex, apex acute, on both sides hirsute; corolla oblique in calyx, yellow, 2.0-2.8 cm long, tube cylindric, 1.8-2.4 cm long, base slightly gibbous, 0.5 cm wide, middle curved, ampliate, throat contracted, 0.5 cm wide, outside glabrous at base, densely yellow-sericeous towards apex, inside pilosulous

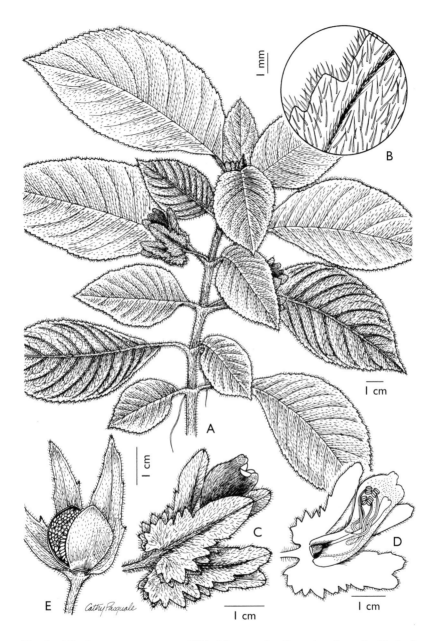

Fig. 1. *Alloplectus savannarum* C.V. Morton: A, flowering branch; B, leaf margin from below; C, calyx and corolla; D, corolla opened to show nectary, stamens, and pistil; E, opened capsule surrounded by persistent calyx. (A-D, Cowan *et al.* 1751; E, Tillett *et al.* 43930).

towards apex, limb 0.3-0.4 cm wide, lobes equal, erect, triangular, 0.2 x 0.2 cm wide, margin nearly entire; stamens adnate to base of corolla tube; ovary ovoid, ca. 0.3 x 0.25 cm, densely pilose, style ca. 2 cm long, glabrous, stigma 2-lobed. Mature capsule white, subglobose, ca. 1.2 x 1 cm.

Distribution: Eastern Colombia to eastern Peru, Brazil (Roraima), Guayana Highlands in Venezuela, Guyana, and Suriname; in disturbed rainforests on mountain slopes, at elevations of 270-1570 m; > 100 collections studied (GU: 50; SU: 6).

Selected specimens: Guyana: Cuyuni-Mazaruni Region: Karowtipu Mt., Boom et al. 7484 (NY, US); Holitipu, Clarke 1072 (CAY, US); Ayanganna plateau, Pipoly et al. 11003 (BBS, CAY, NY, PORT, US); Partang R., Merume Mts., Tillett et al. 43930 (NY, US). Potaro-Siparuni Region: Kaieteur National Park, Gillespie et al. 1263 (B, BPS, CAY, NY, P, US); Mt. Kopinang, Hahn et al. 4287 (US); Mt. Wokomung, Henkel et al. 1248 (BRG, MO, US); Chenapou, Kvist et al. 256 (B, BRG, CAY, US). Suriname: Sipaliwini Distr.: Wilhelmina Mts., Frederick Top, SE of Juliana Top, Holmgren et al. 54396 (NY, NY, US); N of Lucie R., Juliana Top, Irwin et al. 54873 (NY, US).

Phenology: Collected in flower throughout the year.

Note: Clark (2005: 205) placed this species in the Brazilian Atlantic coastal genus *Nematanthus* as *N. savannarum* (C.V. Morton) J.L. Clark.

2. **BESLERIA** L., Sp. Pl. 619. 1753.
   Type: B. lutea L.

Terrestrial or rarely epiphytic, caulescent, erect, rarely scandent herbs or shrubs, without modified stems. Stems rarely branched. Leaves opposite, rarely ternate, equal to unequal in a pair, venation pinnate, foliar nectaries absent. Flowers axillary, rarely solitary, but more usually in few-many fasciculate, epedunculate to long-pedunculate, cymose or subumbellate inflorescences; bracteoles absent; short or long pedicellate. Calyx lobes 5, usually connate at base into a short or long tube; corolla yellow, orange, red, or white, usually cylindric with 5 lobes; stamens included, filaments usually not connate, anthers coherent at first, becoming free, dehiscing by longitudinal slits, thecae parallel to divergent; staminode occasionally developing; disc annular, entire; ovary superior, stigma stomatomorphic to 2-lobed. Fruit a fleshy, indehiscent, red, orange, or white berry.
Chromosome number n=16 (Skog 1984).

Distribution: About 200 species from Mexico to Brazil and the West Indies, from an apparent center in northwestern S America; growing in rainforests, in wet areas or edges of clearings; 7 species in the Guianas.

Notes: Miquel (Linnaea 22: 471. 1849) described *Besleria surinamensis* from Suriname. Pulle (Recueil Trav. Bot. Néerl. 9: 163. 1912) renamed the species as *Besleria verrucosa* (Splitg. ex de Vriese) Pulle based on *Clerodendrum verrucosum* Splitg. ex de Vriese (Ned. Kruidk. Arch. 1: 351. 1848). This species has been identified as *Trichanthera gigantea* (Humb. & Bonpl.) Nees, a member of the ACANTHACEAE.
A single collection of *Besleria lutea* L. is labeled "French Guiana". The specimen at P was collected by L.C. Richard, but the locality is doubtful. The specimen was probably collected in the French Antilles where the species is well known.
Leeuwenberg (1958: 369) reported a collection of *Besleria lanceolata* Urb. in P-LA labeled "French Guiana". Leeuwenberg believed it to have been collected in Martinique where it is endemic.

LITERATURE

Fritsch, K. 1934. Zur Kenntnis der Gattung Besleria. Notizbl. Bot. Gart. Berlin-Dahlem 11: 961-977.

Morton, C.V. 1935. The genus Besleria in British Guiana. Phytologia 1: 151-153.

Morton, C.V. 1935. The genus Besleria in Venezuela. Proc. Biol. Soc. Washington 48: 73-76.

Morton, C.V. 1939. A revision of Besleria. Contr. U.S. Natl. Herb. 26: 395-474.

Morton, C.V. 1968. The Peruvian species of Besleria (Gesneriaceae). Contr. U.S. Natl. Herb. 38: 125-151.

KEY TO THE SPECIES

1 Corolla 3.5-4.0 cm long, oblique in calyx, spurred, pilosulous; peduncle 5-7 cm long; calyx lobes lanceolate, free; ovary pilosulous . . . . *6. B. penduliflora*
Corolla to 2.2 cm long, erect in calyx, not spurred, glabrous to minutely pubescent; peduncle, if present, only up to 4.5 cm long; calyx lobes lanceolate to suborbicular, free or connate to $^2/_3$ of length; ovary glabrous 2

2 Peduncle lacking or much less than 1 cm long 3
Peduncle present, at least 1 cm long 5

3 Leaf margin minutely denticulate to entire; calyx lobes 0.1-0.3 cm long; corolla 0.4-0.7 cm long .... *4. B. parviflora*
Leaf margin serrate, dentate, or sharp serrate towards apex; calyx lobes 0.3-0.8 cm long; corolla 1.2-2 cm long 4

4 Stem apices strigose; calyx lobes connate ²/₃ of length; corolla only slightly longer than calyx, tube not ventricose; disk annular .... *2. B. insolita*
Stem apices hirsute; calyx lobes free; corolla 2-3 x length of calyx lobes, tube slightly ventricose; disk semi-annular .... *7. B. saxicola*

5 Flowers numerous; calyx lobes suborbicular, 0.2-0.3 cm long; corolla small, 0.35-0.5 cm long, white, ventricose .... *1. B. flavovirens*
Flowers 1-8; calyx lobes usually lanceolate, rarely ovate, 0.5-1.5 cm long; corolla 1.3-2.5 cm cm long, variously yellow, orange, or red, but not white, ventricose or not 6

6 Calyx lobes free or rarely connate to ¹/₄ of their length ..... *3. B. laxiflora*
Calyx lobes united for 0.5-0.6 cm .... *5. B. patrisii*

1. **Besleria flavovirens** Nees & Mart., Nova Acta Phys.-Med. Acad. Caes. Leop.-Carol. Nat. Cur. 11: 49. 1823. Type: Brazil, Bahia, Wied-Neuwied s.n. (lectotype BR, here designated, photo US, isolectotypes BR (2)).

Terrestrial herb or subshrub, 0.5-2 m tall. Stem woody at base, erect or ascending, puberulous at apex, glabrescent below. Leaves equal in a pair; petiole 1.4-6 cm long, nearly glabrous; blade papyraceous when dry, oblong lanceolate, 12.5-32 x 5-12 cm, margin serrulate, apex acuminate, base cuneate, above glabrous, below appressed pubescent especially on midrib and veins. Flowers in cymose or congested-paniculate, numerous-flowered inflorescences; peduncle (0.5-)1-2.5 cm long, nearly glabrous; pedicel 0.3-1 cm long, puberulous. Calyx subcampanulate, green or rarely purplish, lobes nearly free, tube ca. 0.1 cm long, free portion of lobes erect, subequal, suborbicular, 0.2-0.3 x 0.2-0.25 cm, margin entire, apex rounded or subemarginate, outside puberulous when young, inside glabrous; corolla erect in calyx, white, 0.35-0.6 cm long, tube cylindric, 0.35-0.4 cm long, base not spurred or gibbous, 0.2 cm wide, middle ventricose, throat contracted, 0.2 cm wide, outside glabrous, inside with a ring of glandular hairs below attachment of filaments, limb 0.3-0.4 cm wide, lobes subequal, spreading, orbicular, 0.1-0.2 x 0.1-0.2 cm, margin entire; stamens subincluded, inserted at middle of corolla tube; ovary broadly ovoid, 0.7 x 0.7 cm wide, glabrous, style 0.3 cm long, glabrous, stigma stomatomorphic. Mature berry reddish, globose-ovoid, 0.4-0.6 x 0.4-0.6 cm.

Distribution: Costa Rica, Colombia (Amazonas), Venezuela (Amazonas, Bolívar), Peru (Loreto), the Guianas, Brazil (Amazonas, Roraima, Río Negro and Bahia); along streams in wet forests, at 40-900 m alt.; > 100 collections studied (GU: 3; SU: 5; FG: 32).

Selected specimens: Guyana: Barima-Waini Region, Matthews Ridge to ridge of "Blue Mt.", McDowell *et al.* 4476 (MO, US); Cuyuni-Mazaruni Region, Kamarang R., Tillett & Tillett 45814 (NY, US); Potaro-Siparuni Region, Pakaraima Mts., Mt. Wokomung, Suruwabaru Cr., Henkel *et al.* 1263 (BRG, MO, US). Suriname: Sipaliwini Distr., near Emmaketen, Daniëls & Jonker 843 (U, US); Nassau Mts., Lanjouw & Lindeman 2372 (K, NY, U, US); Tafelberg, Maguire 24539 (A, NY, U, US). French Guiana: Mont Chauve, Cremers & Crozier 15138 (B, CAY, NY, P, U, ULM, US); Régina Region, Mts. Tortue, Feuillet *et al.* 10110 (AAU, B, BBS, BRG, CAY, COL, E, G, K, L, LE, MEXU, MG, MO, NY, P, PE, U, US, VEN, WAG, WU); Mt. Bellevue de l'Inini, de Granville *et al.* 7812 (B, BR, CAY, INPA, MG, MO, P, U, US, VEN); Mt. Cacao, Skog & Feuillet 5680 (B, BBS, BRG, CANB, CAY, E, KYO, MO, NY, P, PE, SEL, U, UB, US, VEN, WU).

Phenology: Collected in flower and fruit in most months of the year.

Vernacular names: Suriname: gado-oso-tiki (stick of God's house), bergi-tiki (Daniëls & Jonker 843).

2. **Besleria insolita** C.V. Morton, Phytologia 1: 153. 1935. Type: French Guiana, Cayenne, Martin s.n. (holotype K, isotypes BM, FI-W (2)). – Fig. 2

Terrestrial herb, subshrub or shrub, 0.5-2.5(-3) m tall. Stem sappy, erect, apically strigose, glabrescent below. Leaves equal to subequal in a pair; petiole 2.5-11.5 cm long, strigillose; blade papyraceous when dry, elliptic to oblong, 9.5-30 x 4-13 cm, margin serrate to dentate towards apex, apex shortly acuminate, base cuneate, above sparsely strigose, below strigillose especially along midrib and veins. Flowers fasciculate, in 1-6-flowered, epedunculate inflorescences; pedicel (0.5-)1-1.5(-2) cm long, hirto-puberulous. Calyx fusiform-cylindric, yellow, lobes connate about ²/₃ their length, tube 0.7-1.5 cm long, free portion of lobes erect, equal, lanceolate, 0.3-0.6 x 0.1-0.2 cm, margin entire, apex mucronate-bifid, outside hirto-puberulous, inside glabrous; corolla erect in calyx, cream to pale yellow or white, 1.6-2 cm long, tube cylindric, 1.2-2 cm long, base not spurred, 0.3-0.4 cm wide, middle not ventricose, throat slightly constricted, 0.3-0.4 cm wide, outside glabrous, inside with a

14

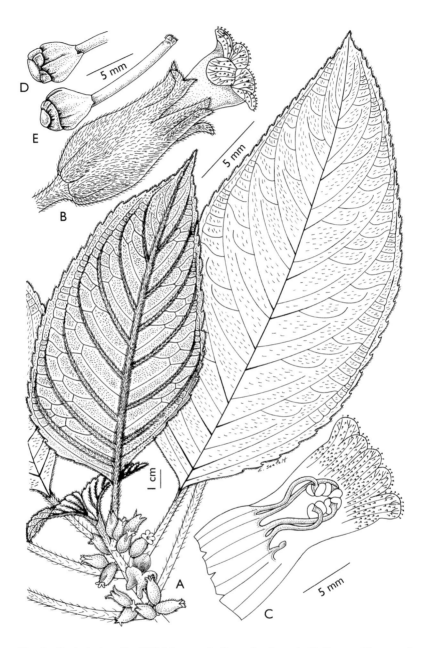

Fig. 2. *Besleria insolita* C.V. Morton: A, flowering branch; B, flower; C, opened corolla with stamens and staminode; D, dorsal side of ovary showing split in nectary; E, ventral side ovary showing style and stigma. (A-E, Skog *et al.* 5682).

ring of glandular hairs in throat, limb 0.6-1 cm wide, lobes subequal, erect to recurved, orbicular to ovoid, 0.3 x 0.2 cm wide, margin entire; stamens included, inserted on corolla tube at 0.4-0.5 cm above base; ovary ovoid, 0.5 x 0.4 cm, glabrous, style 0.7-0.9 cm long, glabrous, stigma 2-lobed. Mature berry red or rarely orange, subglobose, 0.7-1 x 0.6-0.8 cm.

Distribution: Guyana, French Guiana and Brazil (Amapá); locally common along streams in wet forests, at 0-500 m alt.; 59 collections studied, 57 from the Guianas (GU: 1; FG: 56).

Selected specimens: Guyana: Potaro-Siparuni Region, Holmia, Potaro R., Bartlett 8743 (BRG, K, NY). French Guiana, Mts. Tortue, Feuillet *et al.* 10106 (AAU, B, BBS, BRG, CAY, NY, P, PE, U, US, WIS, WU); D.Z. du Haut-Kourcibo, bassin du Sinnamary, de Granville *et al.* 11351 (B, BBS, BR, CAY, G, K, MO, NY, P, U, US); Mt. de la Trinité, NE summit, de Granville *et al.* 6359 (BR, CAR, MO, P, U, US); Mts. de Kaw, Skog & Feuillet 7094 (AAU, BBS, BRG, CAY, G, MO, NY, P, U, US, VEN).

Phenology: Apparently flowering and fruiting throughout the year.

Notes: Hummingbirds observed visiting flowers (Feuillet *et al.* 10106). Photograph: Feuillet & Skog, 2002 (pl. 64 d (Mori & Gracie 21176)).

3. **Besleria laxiflora** Benth., London J. Bot. 5: 361. 1846. Type: Guyana, [Roraima?], Ro. Schomburgk 205.S (holotype K).

Terrestrial suffrutescent herb or low shrub, 0.5-4(-7) m tall. Stem sappy, more or less erect, strigose at apex, glabrescent below. Leaves subequal in a pair; petiole 1.5-7 cm long, sparsely strigillose; blade membranous when dry, elliptic, narrowly oblong, or rarely ovate, 3-21 x 1-8.5 cm, margin shallowly toothed, apex acuminate, base cuneate, above sparsely strigillose on midvein to glabrate, below sparsely strigillose on veins. Flowers umbellate to cymose, in 1- to several-flowered inflorescences; peduncle 1.5-4.5(-6.5) cm long, sparsely strigillose; pedicel 0.8-2.5 cm long, sparsely strigillose. Calyx tubular-campanulate, green to yellow or orange, lobes free or rarely connate to $^1/_4$ their length, tube 0.6-1.7 cm long, free portion of lobes erect, equal, ovate to narrowly lanceolate, 0.5-0.7 x 0.2-0.5 cm, margin entire to minutely fimbriate or ciliate, apex mucronate, outside glabrescent, inside glabrous; corolla erect in calyx, yellow, orange, salmon red, pink to golden red, 1.3-2.3 cm long, tube cylindric, 1.2-2 cm long, base not spurred or gibbous, 0.3-0.7 cm wide, middle slightly ventricose, throat slightly constricted, 0.4-0.7 cm wide, outside glabrous, inside with pubescent ring near attachment of filaments and second ring in throat, limb 0.3-0.5 cm wide, lobes subequal, spreading, ovate, ca. 0.2 x 0.1-0.3 cm wide, margin entire; stamens included, inserted below middle of corolla tube; ovary ovoid, 0.2-0.4 x 0.2-0.3 cm, glabrous, style ca. 0.8 cm long, puberulent, stigma stomatomorphic (in the Guianas). Mature berry orange or red, globose to ovoid, 0.7-1.2 x 1-1.2 cm.

Distribution: Mexico through Central America to Colombia, Venezuela, Brazil, Guyana and Suriname; along streams or wet slopes in dense forests, at 100-1000 m alt.; > 450 specimens studied, 24 from the Guianas (GU: 13; SU: 11).

Selected specimens: Guyana: East Berbice-Corentyne Region, New R., Guppy 370 (= FD 7386) (K, NY); Potaro-Siparuni Region, Iwokrama Mts., Clarke 2525 (US). Upper Takutu-Upper Essequibo Region, NW slopes of Kanuku Mts., Moku-Moku Cr. (Takutu tributary), A.C. Smith 3555 (GH, K, MO, NY, P, U, US). Suriname: Brokopondo Distr., Brownsberg, Stahel & Gonggrijp 19 (= BW 645) (K, U, US), Tjon-Lim-Sang & van de Wiel 32 (= LBB 14771) (BBS, K, U, US); Emmaketen, Gongrijp & Stahel 182 (BBS, U).

Phenology: Collected in flower in October-April, and in fruit in March, September-November.

Vernacular name: Guyana: turuquaréochuru (Guppy 370).

4. **Besleria parviflora** L.E. Skog & Steyerm., Novon 1: 211. 1991.
   Type: Venezuela, Bolívar, Steyermark 74810 (holotype VEN, isotypes F, NY, US). – Fig. 3

Terrestrial suffrutescent herb or shrub, 0.7-4 m tall. Stem woody at base, erect, strigillose to pubescent near apex, glabrescent below. Leaves subequal in a pair; petiole 0.7-5.5 cm long, strigillose to puberulent; blade subcoriaceous to papyraceous when dry, elliptic to lanceolate, 8.5-20.5 x 3.2-7.9 cm, margin minutely denticulate to entire, apex acuminate, base acute to cuneate, above glabrous to sparsely puberulent, below glabrous to sparsely puberulent. Flowers solitary or in 2-8-flowered umbellate inflorescences; peduncle 0-0.4 cm long, puberulous; pedicel 0.2-0.7 cm long, strigillose. Calyx campanulate, green, lobes connate $1/4$ to $1/3$ their length, tube 0.2 cm long, free portion of lobes erect, equal, lanceolate to ovate, 0.1-0.3 x 0.1-0.2 cm wide, margin entire, apex acute to acuminate,

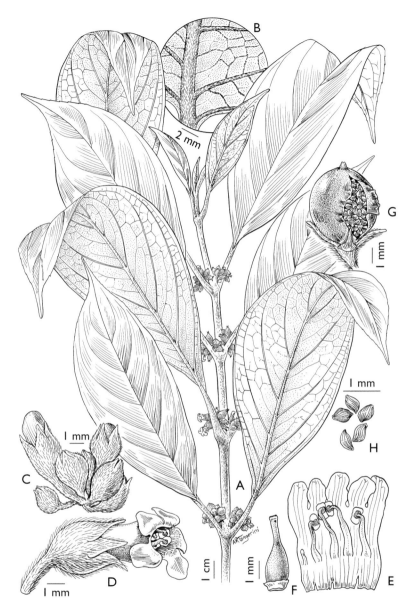

Fig. 3. *Besleria parviflora* L.E. Skog & Steyerm.: A, habit; B, lower leaf surface; C, flower buds; D, flower; E, opened corolla with stamens and staminode; F, pistil and nectary; G, mature berry; H, seeds. (A-F, Steyermark 74810; G-H Steyermark 60052). Reproduced with permission from L.E. Skog & J. Steyermark, Novae Gesneriaceae Neotropicarum III. Additional new species from Venezuela. Novon 1: 211-222. 1991.

outside strigillose, inside glabrous; corolla erect in calyx, greenish white to yellow, 0.4-0.7 cm long, tube cylindric, 0.3-0.6 cm long, base not spurred, 0.15 cm wide, middle not ventricose, throat slightly constricted, 0.2-0.3 cm wide, outside minutely puberulent, inside lacking a ring of glandular hairs, minutely puberulent at mouth, limb 0.3-0.6 cm wide, lobes subequal, slightly spreading, broadly ovate to suborbicular, 0.2-0.25 x 0.2-0.25 cm wide, margin entire; stamens included, inserted ca. 0.15 cm above base of corolla tube; ovary ovoid, ca. 0.2 x 0.12-0.15 cm wide, glabrous, style 0.17-0.2 cm long, glabrous, stigma 2-lobed. Mature berry reddish, globose, 0.3-0.5 x 0.3-0.5 cm.

Distribution: Guayana Highlands in Guyana and adjacent Venezuela and Brazil; along streams in wet forests, at 610 m alt., in Guyana; 18 collections studied (GU: 1).

Specimen studied: Guyana: Potaro-Siparuni Region, Pakaraima Mts., Ireng R. watershed, Manaparu Cr., Mutchnick *et al.* 301 (B, K, MO, NY, U, US).

Phenology: Collected in fruit in October, probably flowering and fruiting throughout the year.

Note: Corolla and stamens not present in Guianan material; descr. from Venezuelan material.

5. **Besleria patrisii** DC., Prodr. 7: 538. 1839. Type: French Guiana, "Cayenne", Patris s.n. (holotype G-DC).

> *Besleria verecunda* C.V. Morton, Phytologia 1: 152. 1935. Type: Guyana, upper Demerara R., Jenman 4156 (holotype US, isotypes K(2), NY, US), syn. nov.
> *Besleria maasii* Wiehler, Selbyana 5: 73. 1978. Type: Cult. Hort. Selby, Wiehler 78123 (holotype SEL, isotypes BH, CAY, K, MO, NY, P, SEL, U, US, VEN), originally from French Guiana, Saül, Maas *et al.* 2281, syn. nov.

Terrestrial herb, subshrub, or shrub, 0.7-2 m tall. Stem sappy, woody at base, erect, sparsely strigose, hirsute, or sericeous at apex, glabrescent below. Leaves equal to subequal in a pair; petiole 0.7-7 cm long, sparsely strigillose, hirsute, sericeous; blade papyraceous or chartaceous when dry, ovate to oblong-elliptic, 5-29 x 1.5-11 cm, margin serrate or repand-serrate, apex acuminate, base cuneate, obtuse to decurrent, above sparsely strigose to sparsely hirsute or pilose, below strigose to hirsute or rarely sericeous along midrib and veins. Flowers solitary, or in umbellate or subcymose 2-8-flowered inflorescences; peduncle 1-4.5(-6) cm long, hirsute to strigillose; pedicel 0.5-3 cm long, sparsely strigillose to sparsely hirsute. Calyx campanulate, yellow to pale green, lobes connate for $^1/_3$ to $^1/_2$ their length, tube 0.5-0.6 cm long, free portion of lobes erect, subequal, lanceolate to oblong-lanceolate, 0.9-1.5 x 0.2-0.5 cm, margin entire, apex subulate-acuminate, outside glabrous to weakly sericeous or sparsely strigose, inside glabrous; corolla erect in calyx, bright orange to red, 1.5-2.5 cm long, tube cylindric, 1.5-2.2 cm long, base not spurred or gibbous, 0.3-0.5 cm wide, middle slightly ventricose, throat slightly constricted, 0.4-0.8 cm wide, outside glabrous, inside with a few hairs to a ring of hairs near attachment of filaments, and a ring of glandular hairs near mouth, limb 0.6-0.9 cm wide, lobes subequal, spreading, broadly ovate to orbicular, 0.2-0.3 x 0.2-0.3 cm, margin entire to obscurely sinuate; stamens included, adnate to below middle of corolla tube to 0.4 cm; ovary ovoid, 0.3-0.4 x 0.25-0.3 cm, glabrous, style ca. 0.8 cm long, glabrous, stigma stomatomorphic. Mature berry red, globose, 0.9-1.2 x 1-1.2 cm.

Distribution: Endemic to the Guianas; on granite boulders or along streams in tall evergreen wet forests, 30-800 m alt.; 85 specimens studied (GU: 4; SU: 13; FG: 68).

Selected specimens: Guyana: Cuyuni-Mazaruni Region, Omai, off "Ridge Road", near Gilt Cr., Gillespie 1508 (B, BBS, K, NY, P, U, US, VEN); Upper Demerara-Berbice Region, Mabura Region, Ekuk compartment, Holder Falls, Ek et al. 714 (US); Sibaruni Cr., left bank Demerara R., Fanshawe 3004 (= FD 6334) (K, NY). Suriname: Sipaliwini, Ouarémapan Cr., Sastre 1773 (CAY, NY, P); E slopes of Bakhuis Mts. between Kabalebo and Coppename Rs., Florschütz & Maas 2890 (MO, NY, U); Frederik Top, 2.5 km SE of Juliana Top, Maguire et al. 54412 (F, K, NY, U, US). French Guiana: Station des Nouragues, Bassin de l'Approuague-Arataye, Nouragues Cr., Cremers 10891 (B, CAY, COL, HAMAB, MPU, NY, P, U, US, VEN); Mt. Bellevue de l'Inini, de Granville et al. 7688 (B, BR, CAY, INPA, MG, MO, P, U, US); Mt. Galbao, E Sect., W slope, de Granville et al. 8663 (B, CAY, E, K, MO, NY, P, U, US, W); Mt. des Singes, 74 km W of Cayenne, near Kourou, Skog et al. 5628 (B, BRG, CAY, K, NY, U, US).

Phenology: Flowering and fruiting throughout the year.

Vernacular name: French Guiana: ampoukou koati (Sastre 6501).

Note: Photograph: Feuillet & Skog, 2002 (pl. 64 e (Mori & Gracie 18726)).

6. **Besleria penduliflora** Fritsch, Repert. Spec. Nov. Regni Veg. 18: 9. 1922. Type: Venezuela, Roraima, Ule 8751 (holotype B destroyed, lectotype G, here designated, isolectotypes F, K, L, U).

Terrestrial shrub, 1.5-6 m tall. Stem woody, erect, internodes tomentose to glabrescent. Leaves unequal in a pair; petiole 0.5-3.2 cm long, tomentose; blade papyraceous when dry, lanceolate to elliptic or oblanceolate, larger blade 3.4-14.0 x 1.5-5.0 cm, margin entire or remotely denticulate, apex acuminate, base cuneate, above sparsely strigillose, below tomentose along veins. Flowers in umbellate or subcymose, 1-3-flowered inflorescences; peduncle (2.5-) 5-7 cm long, tomentose; pedicel 1.5-2 cm long, subtomentose. Calyx subcampanulate, greenish-yellow, lobes nearly free, tube ca. 0.1 cm long, 4 lobes subequal, with dorsal lobe curved around corolla spur and narrower, all lanceolate to oblong, 0.4-0.7 x 0.2-0.4 cm, margin entire, apex acute and recurved, outside strigillose, inside puberulous; corolla strongly oblique in calyx, bright yellow, 3.5-4 cm long, tube cylindric, 3-4 cm long, base with a spur, 0.5 cm wide, middle not ventricose, throat not constricted, 1 cm wide, outside pilosulous, inside with a ring of glandular hairs in throat, limb ca. 1.5 cm wide, lobes subequal, spreading, lobes reflexed, orbicular, 0.4 x ca. 0.4 cm, margin entire; stamens included or slightly exserted, inserted ca. 0.8 cm above corolla base; ovary ovoid, 0.5 x 0.35 cm, densely strigillose to pilosulous, style 3-3.5 cm long, densely strigillose, stigma 2-lobed. Mature berry greenish with red pericarp, globose, 1-1.1 x 1-1.1 cm.

Distribution: Guayana highlands of Venezuela and Guyana; in montane forests, at 915-1600 m alt.; 16 specimens studied (GU: 7).

Selected specimens: Guyana: Cuyuni-Mazaruni Region, NW of northern prow of Mt. Roraima, Hahn *et al.* 5468 (B, US); upper Mazaruni R. basin, Mt. Ayanganna, Tillett & Tillett 45190 (NY, US); Potaro-Siparuni Region, upper Potaro River Region, upper slopes of Mt. Wokomung, Boom & Samuels 9150 (US); Mt. Kopinang, 1350 m, Hahn *et al.* 4299 (NY, US).

Phenology: Collected in flower in February, April, July, and August, fruiting February and July.

Note: *Besleria penduliflora* resembles and is probably closely related to *B. pendula* Hanst. from Venezuela but the latter species differs in having corollas only 1.7-2.4 cm long, calyx lobes pilosulous only at base, and pubescent leaves.

7. **Besleria saxicola** C.V. Morton, Phytologia 1: 151. 1935. Type: Guyana, Potaro R., Tumatumari, Hitchcock 17375 (holotype US, isotypes GH, K, NY, S).

Terrestrial herb, subshrub or shrub, 0.3-1.8 m tall. Stem woody at base, erect, apex hirsute, glabrescent below. Leaves equal to subequal in a pair; petiole 3-12 cm long, hirsute; blade becoming membranous when dry, obliquely elliptic, 20-30 x 7-13 cm, margin remotely sharp-serrulate, apex short-acuminate, base cuneate, above strigillose, below hirsute especially along midrib and veins. Flowers in fasciculate many-flowered, epedunculate inflorescences; pedicel (0.5-)1-1.5 cm long, hirsute. Calyx subcampanulate, yellowish, lobes free, erect, subequal, ovate to ovate-lanceolate, 0.6-0.8 x 0.2-0.3 cm, margin entire and long-ciliate, apex subulate-acuminate, outside hirsute, inside glabrous; corolla oblique in calyx, yellow, 1.8-2 cm long, tube cylindric, 1.7-1.8 cm long, base gibbous, 0.3-0.5 cm wide, middle somewhat ventricose, throat constricted, 0.5-0.6 cm wide, outside glabrous, inside with a ring of glandular hairs in throat, limb 0.7-1 cm wide, lobes subequal, slightly spreading, orbicular, 0.15-0.2 x 0.15-0.2 cm, margin entire; stamens included, inserted below middle of corolla tube; ovary conic, 0.2-0.3 x 0.2-0.3 cm, glabrous, style ca. 1-1.5 cm long, puberulous, stigma 2-lobed. Mature berry yellow, globose, 0.8-1 x 0.8-1 cm.

Distribution: Endemic to Guyana; on rocky slopes and in wet forests, often along streams, 0-610 m alt.; 11 specimens studied (GU: 11).

Selected specimens: Guyana: Barima-Waini Region, between Aranka Head and Barima Head, NW of Kariako R., McDowell *et al.* 4379 (AAU, B, BRG, CAY, COL, K, MO, NY, US, WU); Cuyuni-Mazaruni Region, Kuwara, Kurupung R., Altson 311 (K, NY); Cuyuni-Mazaruni Region, Eping R., McDowell *et al.* 3851 (BBS, BRG, P, U, US); Essequibo R., Moraballi Cr., near Bartica, Sandwith 58 (K, NY); Potaro-Siparuni Region, Tumatumari, Gleason 420 (GH, NY, US).

Phenology: Flowering in January, February, May-July, and October, and fruiting in February, April, May, July, August, and October.

3. **CHRYSOTHEMIS** Decne., Rev. Hort. (Paris) ser. 3. 3: 242. 1849. Type: C. pulchella (Donn ex Sims) Decne. (Besleria pulchella Donn ex Sims)

Terrestrial or rarely epiphytic, caulescent, erect or sometimes decumbent herbs, tuberous. Stems unbranched. Leaves opposite, equal or subequal in a pair, venation pinnate, foliar nectaries absent. Flowers axillary (but appearing terminal by exceeding stem apex), 1-9-flowered cymose or

umbellate, pedunculate inflorescences, with small leafy bracteoles; pedicellate. Calyx lobes 5, basally connate into a tube; corolla yellow to orange, with orange or reddish lines or spots, cylindric, lobes 5; stamens included, filaments basally connate, anthers sometimes coherent, dehiscing by longitudinal slits, thecae parallel; staminode absent; disc a single dorsal, 2-lobed gland; ovary superior, stigma 2-lobed. Fruit a fleshy, brown capsule, loculicidally dehiscent, 2- to several-valved, valves opening slightly.
Chromosome number n=9 (Skog 1984).

Distribution: A genus of 6 species in the West Indies, C and northern S America growing at low elevations; 2 species in the Guianas.

LITERATURE

Moore, H.E. 1954. Chrysothemis and Tussacia. Baileya 2: 86-88.

KEY TO THE SPECIES

1 Leaf blade decurrent into petiole; calyx 5-angled to nearly not angled, the 5
   lobes obvious . . . . *1. C. pulchella*
   Leaf blade not decurrent into petiole; calyx not angled, unlobed, merely
   serrate . . . . *2. C. rupestris*

1. **Chrysothemis pulchella** (Donn ex Sims) Decne., Rev. Hort. (Paris) 21 [ser. 3. 3]: 242. 1849. – *Besleria pulchella* Donn ex Sims, Bot. Mag. 28: ad pl. 1146. 1808. – *Episcia pulchella* (Donn ex Sims) G. Don, Gen. Hist. 4: 656. 1838. – *Tussacia pulchella* (Donn ex Sims) Benth., London J. Bot. 5: 363. 1846. – *Skiophila pulchella* (Donn ex Sims) Hanst., Linnaea 26: 207. 1854 ('1853'). Type: Cult. Woodford, Springwell, Hertfordshire, England (holotype BM).
– Fig. 4

Tussacia villosa Benth., London J. Bot. 5: 363. 1846. – *Chrysothemis villosa* (Benth.) Leeuwenb., Acta Bot. Neerl. 7: 338, fig. 3V. 1958. Type: Guyana, Upper Takutu-Upper Essequibo, Kanuku Mts., Ro. Schomburgk s.n. (holotype K), syn. nov.
Chrysothemis aurantiaca Decne., Rev. Hortic. (Paris) ser. 3. 4: 381. 1850. Type: Cult. Hort. Paris, Aug 1850 (lectotype P) (designated by Leeuwenberg 1958: 335).

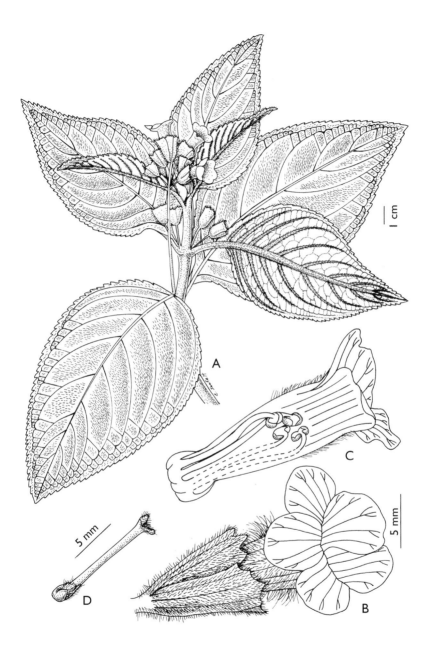

Fig. 4. *Chrysothemis pulchella* (Donn ex Sims) Decne.: A, habit; B, flower; C, corolla showing stamens inside; D, nectary and pistil. (A, A.C. Smith 3579; B-D, Skog & Harvel 4235 [cultivated plant of unknown origin]).

Terrestrial herb, 5-40(-90) cm tall. Stem succulent, erect when young, decumbent with age, densely puberulent to pilose towards apex, glabrescent below. Leaves equal in a pair; petiole <1-3.5 cm long, puberulous to pilose; blade membranous to papyraceous when dry, lanceolate, elliptic to ovate to oblong-ovate, 9-30 x 3-14.5 cm, margin crenate or dentate, apex acute, base short to long decurrent into petiole, above sparsely pilose to subscabrous, or villous, below pilose to puberulous. Flowers in umbellate, 3-9-flowered inflorescences; peduncle <1.5(-5) cm long, puberulous to pilose; pedicel 1.0-2.5 cm long, puberulous to pilose. Calyx tubular-campanulate, not inflated, base rounded, 5-angled or -winged, orange to red, tube 1-1.8 cm long, free portion of lobes erect, subequal, broadly triangular, 0.25-0.5 x 0.35-0.5 cm, margin glandular-denticulate, apex abruptly acuminate, outside pilose to villous, inside glabrous or rarely pubescent; corolla erect in calyx, orange or yellow, 1.6-3 cm long, tube cylindric, base gibbous, 0.3-0.6 cm wide, middle ampliate, throat not constricted, 0.3-0.5 cm wide, outside glabrous near base, white sericeous above, inside glandular pubescent, limb 1-1.5 cm wide, lobes subequal, spreading, suborbicular, 0.3-0.8 x 0.3-0.8 cm, margin entire; stamens included, inserted at base of corolla tube; ovary conic, 0.2-0.4 x 0.2-0.3 cm, pubescent, style 0.7-1.1 cm long, glabrous, stigma deeply 2-lobed. Mature capsule yellow-green, ovoid, ca. 0.6 x 0.6 cm.

Distribution: West Indies, C America (Nicaragua to Panama), Colombia to the Guianas and Brazil (Amazonas, Roraima); usually in damp forests on wet rocks but also on rocky outcrops, at 30-700 m alt.; ± 200 collections studied, 20 in the Guianas (GU: 13; SU: 6; FG: 1).

Selected specimens: Guyana: Cuyuni-Mazaruni Region, Piaitma tipu, McDowell & Gopaul 3051 (B, BBS, CAY, K, NY, P, US); Demerara-Mahaica Region, Great Falls, Demerara R., Jenman 3990 (K, NY, U); Pomeroon-Supenaam Region, Moruka R., de la Cruz 4593 (CM, F, GH, MO, NY, PH, US); Upper Takutu-Upper Essequibo Region, SE Kanuku Mts., Mt. Ishtaban, Gillespie et al. 1888 (B, BBS, CAY, COL, K, MO, NY, P, U, US); Moku-moku Cr. (Takutu trib.), A.C. Smith 3579 (B, F, G, GH, K, MO, NY, P, S, U, US). Suriname: Nickerie Distr., Kabalebo Dam, Lindeman & Görts-van Rijn et al. 323 (BBS, K, NY, U, US, US); Sipaliwini Region, Inselberg Talouakem, Massif de Tumuc-Humac, north face, de Granville et al. 12165 (BBS, CAY, P, US); near Avanavero Falls, Kabalebo R., Corantijn R., Stahel 27 [B.W. 4602] (U, US). French Guiana: Roche Koutou, Bassin du Haut-Marouini, de Granville et al. 9444 (CAY, P, US).

Phenology: Collected in flower in March, June-September.

2. **Chrysothemis rupestris** (Benth.) Leeuwenb., Acta Bot. Neerl. 7: 336, fig. 2, 3R. 1958. – *Tussacia rupestris* Benth., London J. Bot. 5: 363. 1846. Type: Guyana, Kanuku Mts., Ro. Schomburgk s.n. (holotype K).

Terrestrial herb, 5-200 cm tall. Stem succulent, erect, glabrescent, apex puberulous. Leaves equal in a pair; petiole 0.5-5 cm long, puberulous; blade membranous to papyraceous when dry, ovate or oblong-ovate, to 24 x 14.5 cm, margin crenate, apex acuminate, base cuneate or rounded, not decurrent into petiole, above sparsely strigillose, below puberulous, especially on veins. Flowers in 2-7-flowered cymes; peduncle (1-)2-6 cm long, puberulous; pedicel 0.5-2.5 cm long, puberulous. Calyx truncate, infundibuliform, not angled or winged, orange to red, lobes connate into a tube, tube 1-1.5 cm long, free portion of lobes erect, subequal, very small, <0.1 x 0.1-0.2 cm, margin entire, apex acute, outside puberulous, inside glabrous; corolla erect in calyx, red-brown-lined or spotted, 1.8-2.5 cm long, tube cylindric, ca. 1.6 cm long, base somewhat gibbous, 0.2-0.3 cm wide, middle slightly ventricose, throat not constricted, 0.6-0.9 cm wide, outside glabrous below, villous above, inside glandular pubescent, limb 1-1.5 cm wide, lobes subequal, spreading, suborbicular, 0.4-0.5 x 0.4-0.5 cm wide, margin subentire; stamens included, inserted at base of corolla tube; ovary ovoid, 0.2-0.4 x 0.2-0.3 cm, pubescent, style 0.6-0.9 cm long, glabrous, or with a few scattered hairs, stigma 2-lobed. Mature capsule yellow, subglobose, 0.6-0.7 x 0.6-0.7 cm.

Distribution: Endemic to Guianas, but with one unverified specimen from Brazil (Pará), Cid *et al.* 96477; on wet granitic rocks or in rainforests, at 60-950 m alt.; 33 collections studied (GU: 13; SU: 20).

Selected specimens: Guyana: Potaro-Siparuni Region, Iwokrama Mts., Annai-Kurupukari Rd., Hoffman *et al.* 1395 (U, US); Upper Takutu-Upper Essequibo Region, Kanuku Mts., trail to Mt. Iraimakipang, Goodland & Maycock 473 (MTJB, NY); Kanuku Mts., Two-Head Mt., Jansen-Jacobs *et al.* 3497 (BBS, K, U, US). Suriname: Nickerie Distr., Blanche Marie Falls, Maas & Tawjoeran LBB 10961 (BBS, U); Saramacca Distr., Voltzberg, van Donselaar & Schulz LBB 10566 (SEL); Sipaliwini Distr., 9 km N of Lucie R. 12 km W of Oost R., Irwin *et al.* 54558 (NY, US); Augustus Falls, Tafelberg, Maguire 24762 (NY, U, US); confluence of Paloemeu R. and Tapanahoni R., Wessels Boer 1318 (NY, U, US).

Phenology: Collected in flower in February, April, May, and July-September.

4. **CODONANTHE** (Mart.) Hanst., Linnaea 26: 209. 1854 ('1853'),
nom. cons. – *Hypocyrta* Mart. sect. *Codonanthe* Mart., Nov. Gen.
Sp. Pl. 3: 49. 1829.
Type: C. gracilis (Mart.) Hanst. (Hypocyrta gracilis Mart.), typ. cons.

Epiphytic, usually growing on ant-nests, caulescent, pendent, repent, to
erect herbs, shrubs, or lianas, without modified stems. Stems branched or
unbranched. Leaves opposite, equal to rarely strongly unequal in a pair,
venation pinnate, foliar nectaries present on lower leaf surface. Flowers
axillary, solitary or in cymose few-flowered inflorescences;
epedunculate or very short-pedunculate; bracteoles absent; short
pedicellate. Calyx lobes 5, nearly free or briefly connate at base, or
united into 2 lobes (*C. calcarata*); corolla white, pink, lilac, yellow, or
deep purplish, often with reddish lines or spots, funnel-shaped to
subcampanulate, limb 5-lobed; stamens included, filaments basally
connate, anthers coherent in pairs by tips or sides, later separating,
dehiscing by apical pores, thecae separated by broad connectives;
staminode very small; disc a single dorsal, usually large gland; ovary
superior, stigma stomatomorphic to 2-lobed. Fruit a fleshy, indehiscent,
red, pink, orange, or yellow-green berry (in Guianan species).
Chromosome number n=8 or 16 (Skog 1984).

Distribution: A genus of 15 species from Mexico to Brazil, and
from the Lesser Antilles to Peru and Bolivia, on arboreal ant-nests in
rainforests; 2 species in the Guianas.

Note: Plants are usually associated with ants in various ways; growing
from or on arboreal ant-nests, and often with extrafloral nectaries on the
underside of the leaves, between the bases of the calyx lobes, or at the
nodes. The colored seeds are covered by a gelatinous aril or have a
funicle that may be associated with ant dispersal.

LITERATURE

Moore, H.E. 1973. A synopsis of the genus Codonanthe (Gesneriaceae).
Baileya 19: 4-33.

KEY TO THE SPECIES

1 Calyx 2-lobed; leaf margin generally serrulate towards apex . . . .
. . . . *1. C. calcarata*
Calyx 5-lobed; leaf margin entire . . . . *2. C. crassifolia*

1. **Codonanthe calcarata** (Miq.) Hanst., Linnaea 34: 416. 1865 ('1865-1866'). – *Nematanthus calcaratus* Miq., Linnaea 22: 472. 1849. Type: Suriname, Focke 941 (holotype U). – Fig. 5 G

Codonanthe bipartita L.B. Sm., Bull. Torrey Bot. Club 60: 657, f. 1-6. 1933. Type: Guyana, Kartabo, Bailey 181 (holotype GH, isotype NY).

Epiphytic subshrub, usually ca. 30 cm long, occasionally to 1 m. Stem subwoody at base, succulent above, pendent, repent, or creeping, apex puberulent, glabrescent below. Leaves subequal in a pair; petiole rarely up to 1 cm long, puberulent to glabrous; blade coriaceous when dry, oblong-elliptic, 1.5-11.5 cm long, 0.5-4.0 cm wide, margin mostly serrulate near apex, apex acute, base cuneate, above glabrous to minutely puberulent, below glabrous to minutely puberulent. Flowers in cymose, 1-8-flowered inflorescences; epedunculate; pedicel ca. 0.7-1.7 cm long, minutely puberulent. Calyx green, lobes 2, unequal, dorsal lobe free, oblong, 0.3-0.4 x 0.15 cm, margin entire, pushed down by corolla tube, the other lobe ventral, oblong or rectangular, 0.6-0.7 x 0.3-0.4 cm, margin entire, apex rounded and entire or quadridentate, outside and inside puberulous or glabrous; corolla oblique in calyx, white, pinkish, or light purple, sometimes with markings, 1.6-2.5 cm long, tube funnelform, 1.6-2.5 cm long, base spurred, 0.1-0.2 cm wide, middle broader above, throat not constricted, 0.5-1.1 cm wide, outside glabrous, inside sometimes with several hairs near insertion of stamens and minutely glandular-puberulous in throat, limb 0.9-1.6 cm wide, lobes subequal, spreading, suborbicular, 0.3-0.8 x 0.2-0.8 cm, margin entire; stamens included, ventrally adnate for 0.5-0.6 cm to base of corolla tube; ovary ovoid, 0.15-0.2 x 0.1-0.15 cm, minutely puberulous, style 1.5-2 cm long, sparsely pubescent, stigma stomatomorphic. Mature berry purple, blue-black or dark red, globose, ca. 1 x 1 cm.

Distribution: Eastern Venezuela, the Guianas, northern Brazil, and eastern Bolivia; growing on or from arboreal ant-nests in rainforests, at 0-800 m alt.; > 150 collections studied (GU: > 50; SU: 20; FG: 40).

Selected specimens: Guyana: Rupununi Distr., Kuyuwini Landing, Kuyuwini R., Jansen-Jacobs *et al.* 2417 (U); Pakaraima Mts., Kamarang, ca 1 km N of Kamarang, Maas *et al.* 4125 (MO, U, US). Suriname: Lely Mts., SW plateaus, Lindeman & Stoffers *et al.* 647 (U, US); Para R. bank, Went 379 (U, US). French Guiana: between Tonate and Montsinery, Skog & Feuillet 7035 (CAY, NY, UC, US); Piste de Saint-Elie, Billiet & Jadin 1109 (BR, CAY).

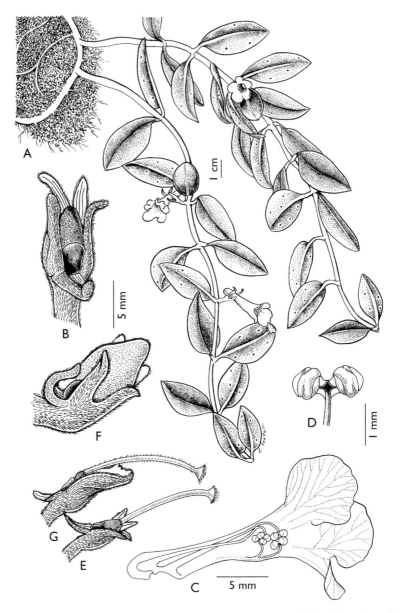

Fig. 5. *Codonanthe crassifolia* (Focke) C.V. Morton: A, habit; B, calyx with nectary and ovary (style removed); C, corolla showing stamens inside; D, apex of stamen showing broad connective; E, calyx showing pistil; F, nectary and fruit in persistent calyx. *Codonanthe calcarata* (Miq.) Hanst.: G, calyx showing pistil. (A, Cremers 5101; B, E, G, Oldeman B-693; C, D, Skog *et al.* 7282; F, Cremers 4232; G, from photo of Feuillet).

Phenology: Flowering and fruiting throughout the year.

Notes: see note under *C. crassifolia*.
Photograph: Feuillet & Skog, 2002 (pl. 65 a (Mori *et al.* 21603)).

2. **Codonanthe crassifolia** (Focke) C.V. Morton, Publ. Field Mus. Nat. Hist., Bot. Ser. 18: 1159. 1938. – *Hypocyrta crassifolia* Focke, Tijdschr. Wis- Natuurk. Wetensch. Eerste Kl. Kon Ned. Inst. Wetensch. 5: 199. 1852. Type: Suriname, Focke s.n. (holotype L).
– Fig. 5 A-F

Codonanthe confusa Sandwith, Bull. Misc. Inform. Kew 1931: 492. 1931. Type: French Guiana, Sagot 426 (holotype K, isotypes BM, BR, P(3), U, W(2)).

Epiphytic subshrub, usually ca. 30 cm long, occasionally to 2 m. Stem subwoody at base, succulent above, pendent, repent, or creeping, apex puberulent, glabrous below. Leaves subequal in a pair; petiole 0.2-1.5 cm long, puberulent; blade coriaceous, often wrinkled when dry, elliptic to oblong or ovate, 1.5-8.5 x 0.6-3.9 cm, margin entire or rarely obscurely sinuate, apex acute to obtuse, base rounded to cuneate, above glabrous to minutely puberulent, below glabrous. Flowers in cymose 1-4(-12)-flowered inflorescences; epedunculate; pedicel 0.5-1.1 cm long, puberulent. Calyx green or reddish, lobes 5, free, dorsal lobe forced back by corolla spur, nearly equal, linear-lanceolate, 0.2-0.9 x 0.1 cm, margin entire, apex acute, outside and inside puberulous; corolla oblique in calyx, white, or cream colored, rarely pink on lobes, 1.5-2.8 cm long, tube funnelform, 1.2-1.8 cm long, base spurred, 0.1-0.2 cm wide, middle broader above, throat not constricted, 0.3-0.8 cm wide, outside glabrous or puberulent, inside with irregular ring of glandular hairs in throat, limb 0.6-1.5 cm wide, lobes subequal, spreading, rotund, 0.2-0.6 x 0.2-0.6 cm wide, margin nearly entire; stamens included, adnate to base of corolla tube for 0.3-0.4 cm; ovary oblong-ovoid, 0.2-0.4 x 0.1-0.2 cm wide, puberulous, style 0.8-1 cm long, glabrous, stigma stomatomorphic. Mature berry pink to red, subglobose, ca. 1 x 1 cm, apex not prominent.

Distribution: As Moore (1973) wrote, "Codonanthe crassifolia has the greatest range and the most variable morphology of any species in the genus". The species is distributed from southern Mexico through C America into northern S America, south to Bolivia and Brazil, and east to the Guianas; growing from or on arboreal ant-nests in rain- or swamp-forests, at 0-700 m alt. (in the Guianas); > 500 collections studied, 110 in the Guianas (GU: 50; SU: 10; FG: 50).

Selected specimens: Rupununi Distr., Kumukowau R., Camp 3, Jansen-Jacobs *et al.* 3853 (BBS, U); Georgetown, Pipoly 7320 (BRG, CAY). Suriname: Nickerie, Area of Kabalebo Dam project, Lindeman & Görts-van Rijn *et al.* 151 (BBS, MO, US); Paramaribo, Agricultural Experiment Station, Maguire & Stahel 22765 (US). French Guiana: Rives de la Comté, env. 5 km S de Roura, Oldeman 1157 (CAY, NY, P, U, US); R. Petite Ouaqui, entre l'ancien village et Saut Verdun, de Granville B-5021 (CAY, P, US).

Phenology: Flowering and fruiting throughout the year.

Notes: Sandwith (Bull. Misc. Inform. Kew 1931: 490-491. 1931) explains that Hanstein's 1865 description of *C. calcarata* was confusing, not based on the type that he apparently had not seen. Hanstein's *C. calcarata* was a different species, which Sandwith described as *C. confusa*, not realizing that the species had already been described as *Hypocyrta crassifolia* by Focke in 1852.
Photograph: Feuillet & Skog, 2002 (pl. 65 c (Mori *et al.* 21610)).

5. **CODONANTHOPSIS** Mansf., Repert. Spec. Nov. Regni Veg. 36: 120. 1934. – *Codonanthe* (Mart.) Hanst. sect. *Codonanthopsis* (Mansf.) H.E. Moore, Baileya 19: 25. 1973.
   Type: C. ulei Mansf.

Epiphytic or epipetric, caulescent, stiffly ascending to erect herbs or subshrubs, without modified stems. Stems branches few. Leaves opposite, strongly unequal in a pair, venation pinnate, foliar nectaries present on lower leaf surface. Flowers sometimes cleistogamous, axillary, solitary or in fasciculate few-flowered inflorescences; epedunculate; bracteoles small; pedicellate. Calyx lobes 5, free; corolla white to yellowish, narrowly funnelform, limb 5-lobed; stamens included, filaments basally connate, anthers coherent in 2 pairs, dehiscing by longitudinal slits, thecae parallel; staminode absent; disc a single dorsal gland; ovary superior, stigma capitate to stomatomorphic. Fruit a fleshy, loculicidally dehiscent, 2-valved, green to purplish capsule, valves recurving to reveal seed mass.
Chromosome number n=9 (Skog 1984).

Distribution: An Andean and Amazonian (from the Guianas to Peru) low elevation rain forest genus of 4 species known from Brazil and Peru to Colombia, east to French Guiana, and in the Guayana region of Venezuela; 1 species in the Guianas.

1. **Codonanthopsis dissimulata** (H.E. Moore) Wiehler, Selbyana 5: 61. 1978. – *Codonanthe dissimulata* H.E. Moore, Baileya 19: 25. 1973. Type: Cult. Bailey Hort., Ithaca, NY (holotype BH), originally from Iquitos, Peru, Stone 1143. – Fig. 6

Epiphytic mostly herbaceous, 30-60(-100) cm long. Stem woody at base, succulent above, ascending and somewhat zig-zag, sparsely to densely pilose towards apex, glabrescent below. Leaves rarely subequal on young shoots, but on mature stems strongly unequal in a pair; petiole 0-0.7 cm long, glabrescent; blade subcoriaceous when dry, lanceolate to elliptic or oblanceolate, subfalcate, blade of larger leaf in a pair 6.5-13.5 x 1.2-5.3 cm, margin entire, ciliate when young, apex acute to acuminate, base subcordate, rounded to acute, occasionally oblique, above glabrous to sparsely pilose, nitid, below glabrous to sparsely pilose along prominent midvein, blade of smaller leaf in a pair bract-like, 0.2-0.5 cm long. Flowers in fasciculate 1-2-flowered inflorescences; pedicel 0.3-1.5 cm long, pilose. Calyx narrowly campanulate, green, lobes free, erect to recurved, unequal, lanceolate, 0.4-1.6 x 0.2-0.6 cm, margin entire, apex acuminate, outside pilose, inside pilose; corolla oblique in calyx, when developing in non-cleistogamous flowers white, ca. 2 cm long, tube narrowly funnelform, 2.4-2.9 cm long, base spurred, 0.2-0.3 cm wide, middle ampliate, subventricose on lower side, throat not contracted, ca. 1 cm wide, outside glabrous at base of tube, sparsely pilose above, inside glandular at base of lower lobe, limb ca. 1.8 cm wide, lobes subequal, spreading, rotund, 0.3-0.4 x 0.4-0.6 cm, margin entire; stamens included, adnate briefly to base of corolla tube; ovary strongly laterally compressed, 0.3 x 0.15-0.2 cm, glabrous to pilose, style 1.2-1.5 cm long, glabrous, stigma capitate to stomatomorphic. Mature capsule green to brown, ellipsoid, 1-2 x 0.5-0.7 cm.

Distribution: Ecuador, northern Peru, Brazil (Amazonas), the Guianas; on fallen trees or large boulders along streams in low wet forests, sometimes on red laterite; 40 collections studied, 4 from the Guianas (GU: 2; FG: 2).

Specimens studied: Guyana: Cuyuni-Mazaruni Region, Omai, Gillespie & Persaud 1596 (BRG, NY, U, US); Potaro-Siparuni Region, Kaieteur Plateau, plane landing to Kaieteur Falls, Cowan & Soderstrom 1837 (SEL, US). French Guiana: near St. Marie les Mines, S of Mt. Cacao, Cremers 6925 (CAY); near Saül, de Granville 2652 (CAY).

Phenology: Collected in flower in December and January, in fruit in January, February and June.

32

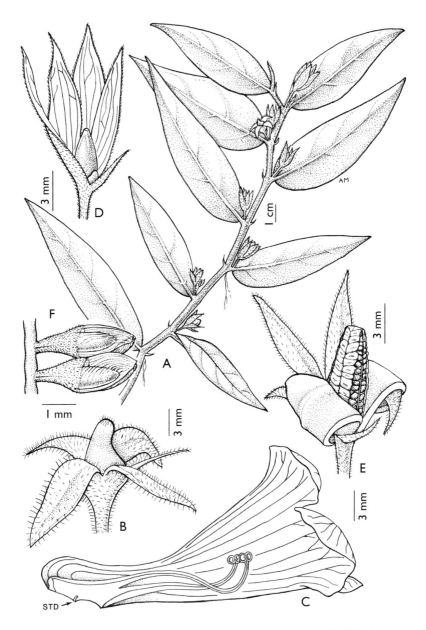

Fig. 6. *Codonanthopsis dissimulata* (H.E. Moore) Wiehler: A, fruiting branch;
B, cleistogamous flower; C, corolla of non-cleistogamous flower opened to
show stamens and staminode [STD]; D, nectary and young fruit in persistent
calyx; E, opened fruit showing valves and seeds; F, seeds. (A, F, Gillespie *et al.*
1596; B, C, E, Dressler 23067 [plant from Peru]; D, Cremers 6925).

6. **COLUMNEA** L., Sp. Pl. 2: 638. 1753.
Type: C. scandens L.

Dalbergaria Tussac, Fl. Antill. 1: 141, pl. 19. 1811-13.
Type: D. phaenicea Tussac [Columnea sanguinea (Pers.) Hanst.]
Trichantha Hook., Icon. Pl. 7: ad pl. 666. 1844.
Type: T. minor Hook. [Columnea minor (Hook.) Hanst.]
Ortholoma (Benth.) Hanst., Linnaea 26: 209. 1854 ('1853').
Type: O. acuminatum (Benth.) Hanst. (Columnea acuminata Benth.)
[Columnea anisophylla DC.]
Pentadenia (Planch.) Hanst., Linnaea 26: 211. 1854 ('1853').
Type: P. aurantiaca (Decne. ex Planch.) Hanst. (Columnea aurantiaca
Decne. ex Planch.) [Columnea strigosa Benth.]

Terrestrial or epiphytic, caulescent, spreading, prostrate, pendulous, to erect herbs or small shrubs, without modified stems. Stems rarely branched. Leaves opposite, equal or unequal in a pair, venation pinnate, foliar nectaries absent. Flowers axillary, solitary or in fasciculate few-10-flowered inflorescences; epedunculate; bracteoles small (in Guianan species), sometimes caducous; usually pedicellate. Calyx lobes 5, usually free nearly to base; corolla usually red, less commonly yellow, rarely greenish or cream, tubular or ventricose, limb 5-lobed; stamens exserted or included, filaments basally connate, anthers coherent in 2 pairs, dehiscing by longitudinal slits, thecae parallel, not divergent; staminode absent or minute; disc a single dorsal 2-lobed gland or rarely 5 separate glands; ovary superior, stigma stomatomorphic or 2-lobed. Fruit a fleshy, indehiscent, white or colored berry.
Chromosome number n=9 (or 18) (Skog 1984).

Distribution: About 160 species from Mexico through C America, to the Caribbean islands and into northern S America as far south as Bolivia and east to Amapá in Brazil; usually in montane rainforests, growing on rocks or epiphytic in trees; 4 species are so far known from the Guianas, a 5th, *C. scandens*, included here because it might be expected in our area.

LITERATURE

Kvist, L.P. & L.E. Skog. 1993. The genus Columnea in Ecuador. Allertonia 6: 327-400.
Morley, B.D. 1972. The distribution and variation of some gesneriads on Caribbean islands. In D.H. Valentine, Taxonomy, phytogeography and evolution. pp. 239-257.

Morley, B.D. 1974. A Revision of the Caribbean species in the genera Columnea L. and Alloplectus Mart. (Gesneriaceae). Proc. Roy. Irish Acad., B 74(24): 411-438.

Morley, B.D. 1976. A key, typification and synonymy of the sections in the genus Columnea L. (Gesneriaceae). Contr. Natl. Bot. Gard. Glasnevin 1(1): 1- 11.

Stearn, W.T. 1969. The Jamaican species of Columnea and Alloplectus. Bull. Brit. Mus. (Nat. Hist.), Bot. 4: 181-236.

KEY TO THE SPECIES

1 Leaves of a pair strongly unequal, the larger leaf in a pair ca. 10 times larger than the smaller 2

Leaves of a pair subequal, the larger leaf in a pair at most 4 times larger than the smaller (*C. calotricha*) 3

2 Leaves with red-purple areas below, sparsely pubescent to pilose above . . . . *2. C. guianensis*

Leaves (in the Guianas) green below, densely tomentose to hirsute above . . . . *4. C. sanguinea*

3 Corolla yellow covered with brown or red hairs, subactinomorphic; leaves as long as to much longer than the flowers . . . . *1. C. calotricha*

Corolla scarlet or red, 2-lipped; leaves usually shorter than the flowers 4

4 Pedicels usually as long as the leaves or longer; leaf blade obtuse to acuminate apically, margin entire, subrevolute, glabrous above; calyx lobes sharply serrate, 1.5-3 times longer than wide . . . . *3. C. oerstediana*

Pedicels usually much shorter than the leaves; leaf blade acute or obtuse apically, margin entire to remotely crenate-dentate, strigose to subtomentose above; calyx lobe base entire or with a few teeth, 2.5-10 times longer than wide . . . . *5. C. scandens*

1. **Columnea calotricha** Donn. Sm., Bot. Gaz. 40: 9. 1905. – *Alloplectus calotrichus* (Donn. Sm.) Stearn, Bull. Brit. Mus. (Nat. Hist.), Bot. 4: 189. 1969. – *Ortholoma calotrichum* (Donn. Sm.) Wiehler, Phytologia 27: 321. 1973. – *Trichantha calotricha* (Donn. Sm.) Wiehler, Selbyana 1: 34. 1975. Type: Guatemala, Türckheim 8542 (holotype US, isotype US).   – Fig. 7 A-G

Columnea calotricha Donn. Sm. var. *austroamericana* C.V. Morton, Bol. Soc. Venez. Ci. Nat. 23: 78. 1962. Type: Suriname, Brownsberg, Stahel & Gonggrijp 122 (= BW 626) (holotype US, isotype U), syn. nov.

Columnea calotricha Donn. Sm. var. *breviflora* C.V. Morton, Bol. Soc. Venez. Ci. Nat. 23: 78. 1962. Type: Costa Rica, J. Donnell Smith 6728 (holotype US), syn. nov.

Epiphytic or rarely terrestrial herb or subshrub, 0.1-1 m tall. Stem succulent, erect or ascending, hirsute. Leaves subequal or rarely strongly unequal in a pair; petiole 0.5-3 cm long, pilose; blade membranous to papyraceous when dry, lanceolate, elliptic or obovate, larger blade 3-19 x 1-6 cm, margin serrulate or crenulate, apex acute to acuminate, base cuneate, usually oblique, above and below pilose. Flowers solitary or up to 7 in fasciculate inflorescences; pedicel 0.3-1 cm long, hirsute. Calyx subcampanulate, green at base, red or orange above, lobes free, erect, subequal, oblanceolate or oblong, 0.8-2.5 x 0.3-0.6 cm, margin serrate near apex, apex acute and serrulate, outside densely pilose to hirsute, inside hirsute towards apex; corolla erect in calyx, orange-yellow or yellow with brown or red hairs, 1.5-3.5 cm long, tube narrowly tubular, 1.5-2.0(-3.5) cm long, base gibbous, 0.2-0.3 cm wide, middle only slightly curved, not ventricose, throat slightly contracted, 0.2-0.5 cm wide, outside brown-pilose, inside glabrous, corolla limb 0.6-1 cm wide, lobes subequal, erect, triangular to oblong, 0.3-0.4 x 0.2-0.3 cm, dorsal lobes concave, margin entire; stamens equalling corolla tube to exserted, adnate to corolla tube base; staminode minute; ovary ovoid, 0.3 x 0.2 cm, sparsely hirsute, style 1.7-3.5 cm long, glabrous, stigma 2-lobed. Mature berry white below, purplish to reddish above, ovoid, ellipsoid to globose, 0.5-1(-1.6) x 0.4-0.5 cm.

Distribution: Guatemala, Costa Rica, Panama, Brazil (Amapá), in Suriname and French Guiana; growing in moss on trees, in swamp and rain forests, at 0-750 m alt.; > 100 specimens studied, 63 from the Guianas (SU: 11; FG: 52).

Selected specimens: Suriname: Brokopondo Distr., Tawjoeran LBB 12558 (BBS, US); Brownsberg, Zaandam s.n. (= BW 6619) (K, U, US); Sipaliwini Distr., Emmaketen, Stahel 15 (= BW 5723) (U, US). French Guiana: région de Paul Isnard, Mt. Lucifer, Feuillet 352 (CAY, P, US); Tamouri, de Granville 2099 (CAY, P, US); Saut Mapaou, Approuague R., Oldeman B-552 (CAY, MO, NY, P, U, US).

Phenology: Collected in flower in all months of the year, in fruit in April.

Vernacular names: Suriname: aloekoe wonoeloa (Car.; Stahel & Gonggrijp 13); jaloealoea (Car.); yamul ka a (Wayampi). French Guiana: suisuika'a (Wayãpi; Grenand 310), suwisuwika'a (Wayãpi, Prévost & Grenand 2013); yamulepila (Wayãpi; Jacquemin 1546).

2. **Columnea guianensis** C.V. Morton in Maguire *et al.*, Bull. Torrey Bot. Club 75: 564. 1948. – *Alloplectus guianensis* (C.V. Morton) Stearn, Bull. Brit. Mus. (Nat. Hist.), Bot. 4: 189. 1969. – *Dalbergaria guianensis* (C.V. Morton) Wiehler, Phytologia 27: 317. 1973. Type: Guyana, Maguire & Fanshawe 23067 (holotype US, isotypes A, F, K, MO, NY, U, VEN). – Fig. 7 H

Epiphytic or epipetric large herb or small shrub, to 3.2 m tall. Stem woody at base, succulent above, procumbent and ascending, villous or hirsute to glabrescent. Leaves strongly unequal in a pair; petiole 0.4-1 cm long, villous or hirsute; blade chartaceous when dry, oblanceolate, blade of larger leaf in a pair 12-28.5 x 4.3-10.9 cm, margin denticulate, apex acute to acuminate, base obliquely cuneate, above sparsely pilose to glabrous, below pubescent to pilose. Flowers solitary or up to 3 in fasciculate inflorescences; pedicel 0.3-1 cm long, villous. Calyx campanulate, green, lobes nearly free, erect, subequal, lanceolate, narrowed towards base, 2-2.6 x 0.5-0.6 cm wide, margin sharply serrate, apex long-acuminate, outside villous, sometimes with red hairs, inside villous; corolla erect in calyx, pale yellow with reddish lobes, 2.5-3.3 cm long, tube narrowly cylindric, 2-3.2 cm long, base subgibbous, 0.4-0.5 cm wide, middle ventricose, throat subgibbous, 0.4-0.6 cm wide, outside sericeous, inside sparsely puberulous, limb ca. 0.8 cm wide, lobes subequal, spreading, subtriangular, ca. 0.2 x 0.2 cm wide, margin reddish and entire; stamens included, adnate to corolla base; staminode very small; ovary ovoid, ca. 0.4 x 0.3 cm, pubescent, style ca. 2.5 cm long, glabrous, stigma 2-lobed. Mature berry white, ellipsoid, 1 x 0.6 cm.

Distribution: Guyana and adjacent Venezuela (Sucre); wet forests often among rocks, at 180-1200 m alt.; 15 collections studied, 12 in the Guianas (GU: 12).

Selected specimens: Guyana: Cuyuni-Mazaruni Region: Chi-Chi Mts., W of Chi-Chi Falls, Pipoly *et al.* 10275 (FDG, MO, NY, U, US), Upper Mazaruni R. basin, Kamarang R., Bailey line to Karowtipu, Tillett & Tillett 45483 (NY); Potaro-Siparuni Region: Upper Potaro R. Region, ca. 19 mile N of Kopinang village, Boom & Samuels 8910 (US), between Kaieteur Falls and Tukeit, Kvist *et al.* 214 (BRG, U, US), Kaieteur Falls, Kvist *et al.* 364 (AAU, B, BRG, COL, US).

Phenology: Collected in flower in February, April, June, July, September and October.

3. **Columnea oerstediana** Klotzsch ex Oerst., Centralamer. Gesner. 61, pl. 8. 1858. Type: Costa Rica, Oersted 9291 (holotype C, isotypes C, US).

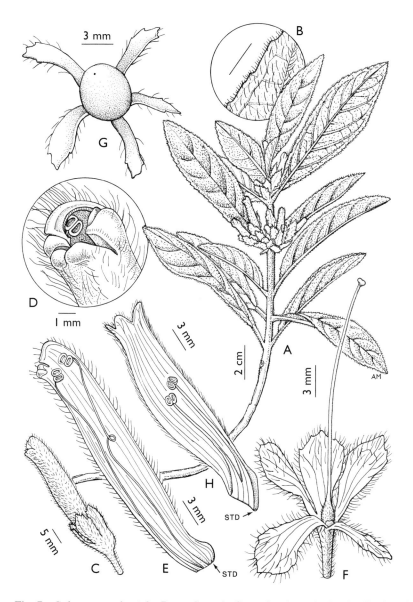

Fig. 7. *Columnea calotricha* Donn Sm.: A, flowering branch; B, detail of leaf margin from below; C, flower; D, mouth of corolla showing anthers; E, corolla opened to show stamens and staminode [STD]; F, flower with corolla and stamens removed to show nectary and pistil; G, berry. *Columnea guianensis* C.V. Morton: H, corolla opened and showing stamens and staminode [STD]. (A-B, [habit and leaves] Oldeman 2959, [inflorescence] Feuillet 4381; C-F, Cremers 9903; G, Prévost 1885; H, Maguire & Fanshawe 23067).

Epiphytic suffrutescent herb or subshrub, to 1 m long. Stem woody at base, succulent above, pendent, creeping or erect, pubescent or strigose when young. Leaves equal in a pair or whorl; petiole 0.1-0.2 cm long, pubescent; blade chartaceous or coriaceous when dry, ovate, elliptic-ovate, to oblong, 1.2-2.9(-3.7) x 0.2-1.2 cm, margin entire, subrevolute, apex obtuse to acuminate, base rounded to cuneate, above glabrous, below sparsely strigose. Flowers solitary in reduced fasciculate inflorescences; pedicel 1-4 cm long, pubescent to strigose. Calyx campanulate, green to red, lobes free, erect, subequal, ovate-lanceolate, 0.9-1.8 x 0.4-1.1 cm, margin toothed near base or nearly entire, apex long acuminate, outside pubescent, inside sparsely pubescent; corolla erect in calyx, scarlet or red, 2.5-5.5 cm long, tube funnelform, 1.5-2.5 cm long, base gibbous, above base 0.2-0.4 cm wide, middle slightly ventricose, throat not contracted, 0.7-0.9 cm wide, outside pilose, inside glabrous, limb ca. 3 cm wide, lobes strongly unequal, upper lobes connate into a galea, erect, 1-2 x 0.7-1.5 cm, margin entire, lateral lobes spreading, triangular, 0.5-1.1 x 0.5-1 cm [at base], margin entire, basal lobe reflexed, triangular, 1.2-1.6 x 0.3-0.4 cm wide, margin entire; stamens exserted, adnate to base of corolla tube; staminode absent; ovary ovoid, 0.3-0.4 x 0.2-0.3 cm, pubescent to sericeous, style ca. 5 cm long, pilose glandular, stigma 2-lobed. Mature berry white, subglobose, 0.6-0.8 x 0.6-0.8 cm.

Distribution: Costa Rica, Panama, French Guiana and Brazil (Amapá); pendent from trees in wet forests, at 0-760 m alt.; ± 150 collections studied, 58 from the Guianas (FG: 58).

Selected specimens: French Guiana: Sommet Tabulaire, 50 km SE de Saül, Cremers 6420 (CAY, P, U, US); R. Kourcibo, Cremers & Pawilowski 13730 (B, BR, CAY, G, K, MO, NY, P, U, US); Région de Paul Isnard, Mt. Lucifer, Feuillet 328 (CAY, P, US); région de Régina, Mts. Tortue, 11 km WNW de Approuague R., Feuillet et al. 10231 (B, BBS, CAY, BRG, P, U, US, WAG); R. Arataye, Saut Pararé, de Granville 8 (CAY, NY, P, U); Mt. Galbao, SW de Saül, de Granville 2364 (CAY, P, U); Comté R., de Granville B-4666 (CAY, P, U, US).

Phenology: Flowering throughout the year.

4. **Columnea sanguinea** (Pers.) Hanst., Linnaea 34: 384. 1865 ('1865-1866'). – *Besleria sanguinea* Pers., Syn. Pl. 2: 165. 1807. – *Alloplectus sanguineus* (Pers.) G. Don, Gen. Hist. 4: 655. 1838. – *Dalbergaria sanguinea* (Pers.) Steud., Nomencl. Bot. ed. 2. 1: 479. 1840. – *Collandra sanguinea* (Pers.) Griseb., Pl. Wright. 526. 1862. Type: Hispaniola: Turpin (not located).

Columnea aureonitens Hook., Bot. Mag. 73: ad pl. 4294. 1847. – *Collandra aureonitens* (Hook.) Hanst., Linnaea 26: 209. 1854 ('1853'). – *Dalbergaria aureonitens* (Hook.) Wiehler, Phytologia 27: 316. 1973. Type: [icon] Hook., Bot. Mag.73. pl. 4294. 1847 (neotype, designated by Leeuwenberg 1958: 383), syn. nov.

Terrestrial or epiphytic subshrub, to 2 m tall, rarely taller. Stem woody at base, succulent above, erect or ascending, tomentose to hirsute at apex, becoming glabrescent below. Leaves strongly unequal in a pair; petiole 0.4-1.8 cm long, tomentose; blade papyraceous when dry, oblanceolate, larger blade 15.5-32.6 x 5.4-10.7 cm, margin serrate to dentate, apex acuminate, base obliquely cuneate, above hirsute to strigose, hairs transparent but occasionally reddish towards margins, below tomentose to hirsute, hairs transparent. Flowers solitary, or up to several in fasciculate inflorescences; pedicel 0.2-0.5 cm long, reddish hirsute. Calyx campanulate, greenish-white to orange or scarlet, lobes nearly free, erect, subequal, lanceolate, 1.3-2.1 x 0.3-0.6 cm, margin laciniate or 3-6-toothed, apex acuminate to obtuse, outside whitish or reddish hirsute, inside hirsute; corolla slightly oblique in calyx, yellow, 2-3 cm long, tube cylindric, 1.8-2.7 cm long, base subgibbous, above base ca. 0.4 cm wide, middle slightly ventricose, throat slightly contracted, 0.5-0.6 cm wide, outside reddish to golden hirsute, inside pubescent, limb 0.5-0.6 cm wide, lobes subequal, erect, ovate, 0.2-0.4 x 0.2-0.3 cm, margin entire; stamens included, adnate to base of corolla tube; staminode very small; ovary ovoid, 0.4-0.5 x 0.2-0.3 cm, hirsute to pilose, style ca. 2.5 cm long, glabrous, stigma shortly 2-lobed. Mature berry white to orange, subglobose, ca. 1.8 x 1 cm.

Distribution: C America, northern S America to Ecuador and Bolivia, in Suriname and French Guiana, and the West Indies; growing in forests, often on mossy tree trunks, at 150-1250 m alt. (in the Guianas); ± 350 collections examined, 36 from the Guianas (SU: 4; FG: 32).

Selected specimens: Suriname: Sipaliwini Distr., near Julianatop, Irwin *et al.* 54766 (NY, US), 55100 (NY, US); Wilhelmina Mts., Stahel 438 (= BW 7059) (U). French Guiana: Mt. de l'Inini, Feuillet 3727 (CAY, US); Sommet Tabulaire, 50 km SE de Saül, de Granville 3503 (CAY, P, U, US); Mt. Galbao, WSW of Saül, Leeuwenberg 11736 (CAY, NY, P); R. Yaroupi, Saut Tainous, Oldeman 3127 (CAY, P, U, US).

Phenology: Collected in flower February-October.

Vernacular names: suwisuwika'awa (Wayãpi); kaleaku away (Wayãpi; de Granville 2474); yamul ka a (Wayãpi; Jacquemin 1698).

Notes: The Guianan plants of this species have traditionally been placed in *Columnea aureonitens* Hook. The characters distinguishing that species from *C. sanguinea* were the color and investiture of the lower leaf surface. The plants of the two taxa display all the characteristics and intermediates that cannot be distinguished reliably. The typical form of this species has prominent red areas near the apices on the lower leaf surfaces as seen in populations from Hispaniola, Trinidad and elsewhere. The all-green form of the species is found in Cuba, C America to Ecuador and Bolivia, and the Guianas.

5. **Columnea scandens** L., Sp. Pl. 638. 1753. Type: [icon] Plumier, Pl. Amer. pl. 139, fig. 1. 1756 (neotype, designated by Leeuwenberg 1958: 390).

Terrestrial or epiphytic subshrub, 0.3-0.5 m long or more. Stem succulent, climbing or pendent, sarmentose, sericeous-strigose near apex, becoming glabrous below. Leaves equal to subequal in a pair; petiole 0.2-1.1 cm long, sericeous-strigose; blade papyraceous or chartaceous when dry, elliptic or oblong-elliptic, 1.5-6.5 x 0.9-3.8 cm, margin entire to remotely crenate-dentate, apex acute or obtuse, base cuneate or rounded, above glabrous to strigose or subtomentose, below glabrous to strigose or subtomentose. Flowers in fasciculate 1-3-flowered inflorescences; pedicel 0.8-2 cm long, sericeous-strigose. Calyx campanulate, green or reddish, lobes free or lateral and ventral lobes connate 0.1-0.2 cm, tube 0.1-0.2 cm long, free portion of lobes erect, subequal, ovate-lanceolate to linear, 0.8-1.7 x 0.1-0.4 cm, margin entire at base or with a few teeth, apex acuminate, subulate or acute, outside strigose, inside strigose; corolla erect in calyx, red, 4.3-7 cm long, tube cylindric, 2.3-4.9 cm long, base gibbous, 0.15-0.4 cm wide, middle slightly ampliate, throat not contracted, 0.5-0.9 cm wide, outside pilosulous, inside sparsely pubescent on dorsal side, limb ca. 2.5 cm wide, lobes strongly unequal, upper lobes connate into a galea, erect, 1.3-1.8 x 1-1.1 cm, margin entire, lateral lobes spreading, triangular, 0.7-0.9 x 0.7-1 cm, margin entire, basal lobe reflexed, lanceolate, 1.2-1.5 x 0.25-0.4 cm, margin entire; stamens exserted, adnate to base of corolla tube; staminode absent; ovary ovoid, 0.3-0.35 x 0.15-0.25 cm, appressed-pubescent or glabrous, style ca. 4.5 cm long, pubescent, stigma 2-lobed. Mature berry white, globose, ca. 0.8 x 0.8 cm.

Distribution: West Indies and northern S America; in rainforests, at low to middle elevations.

Note: The description above is compiled from that given by Leeuwenberg (1958, 1984) as we have seen no authentic specimens of *Columnea scandens* from the Guianas. The species may yet be found there for it is known from Venezuela, Trinidad and the Lesser Antilles. Leeuwenberg (1958) cited two collections from the Guianas, the first "Cayenne, Martin 100 (BM, p.p.)", may well have come from Martinique where Martin also collected; the second, a specimen cited by Richard Schomburgk (1849: 972) as collected by Robert Schomburgk in 1841 from "near Aruka" (probably the Aruka R.) in Guyana, was in the Berlin Herbarium and now no longer extant. No duplicates of this latter specimen have been found.

7. **CORYTOPLECTUS** Oerst., Centralamer. Gesner. 45. 1858.
   Type: C. capitatus (Hook.) Wiehler (Alloplectus capitatus Hook.)

Terrestrial, caulescent, erect herbs, without modified stems. Stems unbranched. Leaves opposite, nearly equal in a pair, venation pinnate, foliar nectaries absent. Flowers axillary, 1 to few in umbellate-cymose inflorescences; short pedunculate; bracteoles small; long pedicellate. Calyx lobes 5, free; corolla yellowish, tubular, inflated ventrally, limb 5-lobed; stamens included, filaments basally connate, anthers free or coherent in 2 pairs, dehiscing by longitudinal slits, thecae parallel; staminode present, small; disc of 2 or 4 opposite separate glands, or a double connate dorsal gland; ovary superior, stigma capitate or 2-lobed. Fruit a fleshy, indehiscent, black berry.
Chromosome number n=9 (Skog 1984).

Distribution: An Andean genus of 8 or more species, in mixed wet or cloud forest, known from Bolivia to coastal Venezuela, and in the Guayana Highlands of Venezuela (Bolívar, Amazonas), neighbouring Brazil, and western Guyana; 1 species in the Guianas.

1. **Corytoplectus deltoideus** (C.V. Morton) Wiehler, Phytologia 27: 313. 1973. – *Alloplectus deltoideus* C.V. Morton, Fieldiana, Bot. 28: 521. 1953. Type: Venezuela, Bolívar, Steyermark 60170 (holotype US, isotypes F, VEN).   – Fig. 8

Terrestrial herb, 0.6-1.5 m tall. Stem woody at base, succulent above, erect, apex densely velutinous pubescent with reddish glandular hairs. Leaves subequal in a pair; petiole 3-7.5 cm long, appressed pilose; blade chartaceous when dry, elliptic, 11-22 x 4.5-8.9 cm, margin crenulate to subserrulate, apex acuminate, base acute to oblique, above scabridulous, below sparsely to densely strigillose. Flowers 2-3 in fasciculate inflorescences; peduncle 0-0.2 cm long, pubescent; pedicel 2.5-5.5 cm

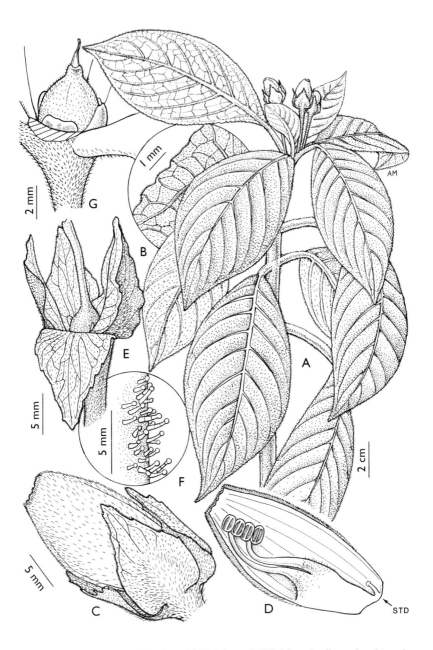

Fig. 8. *Corytoplectus deltoideus* (C.V. Morton) Wiehler: A, flowering branch; B, lower leaf surface and margin; C, flower; D, corolla opened to show stamens and staminode [STD]; E, flower with corolla removed to show pistil; F, detail of glandular hairs on pedicel; G, young fruit. (A-G, Renz 14223).

long, subsericeous-velutinous. Calyx broadly campanulate, green to red, lobes free, erect, subequal, broadly deltoid, 1.1-1.9 x 0.6-1.3 cm, margin entire, apex long acuminate, outside glabrescent to densely strigillose, inside glabrescent to densely strigillose; corolla erect in calyx, yellow, ca. 2 cm long, tube subventricose, 1.8-1.9 cm long, base gibbous, 0.45-0.6 cm wide, middle slightly curved, throat contracted, ca. 0.5 cm wide, outside villous, inside glabrous, limb ca. 0.5 cm wide, lobes subequal, erect, subrotund, ca. 0.05 x ca. 0.1 cm, margin entire; stamens included, attached briefly to corolla base; ovary broadly ovoid, 0.3-0.5 x 0.3 cm, pilose, style ca. 2.8 cm long, pilose, stigma 2-lobed. Only immature fruit seen, subglobose.

Distribution: Venezuela (Bolívar) and Guyana; in high mixed forests, among mosses on forest floor with sandstone substrate near streams, at 700-2250 m alt. (1300 m in the Guianas, based on insufficient specimen data); 10 collections studied (GU: 4).

Specimens studied: Guyana: Cuyuni-Mazaruni Region: Waukauyengtipu, Clarke *et al.* 5526 (US); Mt. Roraima-Waruma trail, Persaud 84 (K); E bank of Waruma R., 20 km S of confluence with Kako R., Renz 14169 (U); N slope of Mt. Roraima, Renz 14223 (U).

Phenology: Collected in flower in February, July, and October, probably fruiting in October or later.

8. **CREMERSIA** Feuillet & L.E. Skog, Brittonia 54: 347. 2003 ('2002').
    Type: C. platula Feuillet & L.E. Skog

Terrestrial, caulescent, erect, small herbs, without modified stems. Stems unbranched. Leaves opposite, equal or subequal in a pair, venation pinnate, foliar nectaries absent. Flowers axillary, in racemose few-flowered inflorescences; pedunculate; bracteoles small; pedicellate. Calyx lobes 5, free; corolla pale purple, salverform, limb 5-lobed; stamens included, filaments basally connate, anthers apically coherent in a tetrad, dehiscing by longitudinal slits, thecae parallel; staminode minute; disc a single dorsal gland; ovary superior, stigma capitate. Fruit a dry, loculicidally dehiscent, 2-valved, yellowish-green capsule, valves opening to 180°.
Chromosome number unknown.

Distribution: Monotypic, one species endemic to southern French Guiana, in rain forest, on granitic rocks or boulders.

44

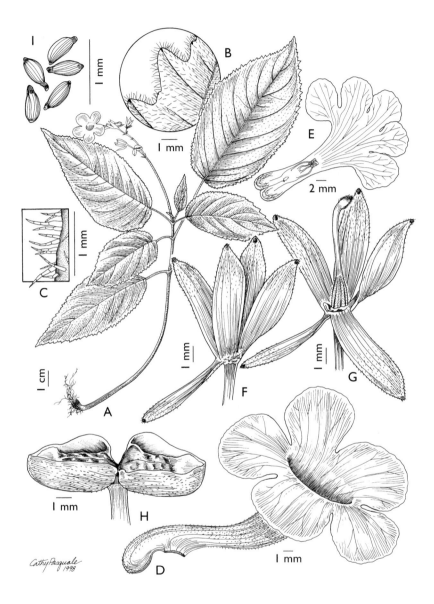

Fig. 9. *Cremersia platula* Feuillet & L.E. Skog: A, habit; B, leaf margin; C, hairs on leaf; D, corolla; E, corolla opened to show stamens and staminode; F, calyx; G, calyx, with one sepal bent to show nectary and pistil; H. capsule (calyx removed); I, seeds. (A-I, Cremers 13126).

1. **Cremersia platula** Feuillet & L.E. Skog, Brittonia 54: 348. 2003 ('2002'). Type: French Guiana, Mt. Bakra, Cremers 13126 (holotype US, isotypes B, CAY, NY, P, U). – Fig. 9

Terrestrial herb, 10-30 cm tall. Stem thin, erect, pilose. Leaves subequal in a pair; petiole 1.5-4 cm long, pilose; blade papyraceous or chartaceous when dry, ovate to elliptic, 5-11 x 2.5-5 cm, margin serrulate to biserrulate, apex acute to obscurely acuminate, base round to slightly cordate and often unequal, above with scattered hairs, below with scattered hairs. Flowers in pseudoraceme-like cymose 3-15-flowered inflorescences; peduncle ca. 7 cm long, pilose; pedicel 1-1.5 cm long, pilose. Calyx green, reddish in bud, 4 lobes free, erect, equal, linear-lanceolate, 0.7-0.8 x 0.22 cm, the 5th one dorsal, recurved, margin entire, apex blunt, glandular, outside and inside pilose in upper half; corolla transversal-oblique in calyx, pale purple, 2-2.3 cm long, tube salverform, 1.5-1.8 cm long, base spurred, 0.25 cm wide, middle slightly curved, broadening towards throat, throat narrower dorsally, 0.85 cm wide, outside pilose, inside glabrous, limb 2.2-2.7 cm wide, lobes subequal, spreading, suborbicular, 0.7-0.8 x 0.7-0.8 cm, margin nearly entire; stamens included, adnate to tube in basal third; ovary conical, 0.2-0.22 x 0.1 cm (at base), pilose, style 0.4-0.43 cm long, glabrous, stigma capitate. Mature capsule brown, ovoid, 0.5-0.6 x 0.2-0.25 cm.

Distribution: Endemic to southern French Guiana, Mt. Bakra, under the canopy of the rainforest, on granitic cliff bases (de Granville 14868) or boulders (Cremers 13126); known from only 2 collections from the same locality in French Guiana.

Specimens studied: French Guiana: Mt. Bakra, Cremers 13126 (B, CAY, NY, P, U, US), de Granville 14868 (CAY).

Phenology: Collected in flower and fruit in April and June.

9. **DRYMONIA** Mart., Nov. Gen. Sp. Pl. 3: 57. 1829.
   Type: D. calcarata Mart. [Drymonia serrulata (Jacq.) Mart.]

Terrestrial or epiphytic, caulescent, creeping or climbing, sometimes erect herbs, shrubs, or lianas, without modified stems. Stems branched or unbranched. Leaves opposite, equal or nearly equal in a pair, venation pinnate, foliar nectaries absent. Flowers axillary, solitary, or in fasciculate several-flowered inflorescences; pedunculate or epedunculate; bracteoles often caducous, sometimes absent; pedicellate. Calyx lobes 5, free or briefly connate at base; corolla white, yellow, purple to red, often with markings, usually funnelform and broader towards mouth, limb 5-lobed; stamens included, filaments basally connate, anthers at first coherent by sides and faces, later separating, dehiscing by short basal slits, thecae divergent at base; staminode small to minute; disc a single dorsal gland; ovary superior, stigma stomatomorphic or 2-lobed. Fruit a fleshy capsule, becoming coriaceous, interior red, orange, or purple, loculicidally dehiscent, 2 valves recurving and spreading widely.

Chromosome number n=9 (Skog 1984).

Distribution: More than 200 species throughout the continental Neotropics and the Lesser Antilles, from a probable center of distribution in Colombia and Ecuador; in moist forests; 4 species in the Guianas.

KEY TO THE SPECIES

1 Leaves of a pair strongly unequal . . . . *3. D. psilocalyx*
  Leaves of a pair subequal 2

2 Bracts lacking (flowers solitary or 1-3) . . . . *4. D. serrulata*
  Bracts present 3

3 Bracts white, spotted with red, smaller than calyx lobes . . *1. D. antherocycla*
  Bracts pink to dark red, larger than calyx lobes . . . . *2. D. coccinea*

1. **Drymonia antherocycla** Leeuwenb., Acta Bot. Neerl. 14: 155, fig. 1. 1965. Type: Suriname, Wessels Boer 1112 (holotype U, isotypes NY, US).

Epiphytic herb, 75 cm high. Stem subwoody at base, succulent above, erect?, appressed-pubescent at apex. Leaves subequal in a pair; petiole 1-5 cm long, sparsely appressed-pubescent; blade papyraceous to chartaceous when dry, narrowly elliptic, 10-40 x 4-11 cm, margin obscurely dentate, apex acuminate, base obliquely long-cuneate, above sparsely appressed-pubescent, below more densely appressed pubescent. Flowers solitary to several; epedunculate; bracts white, spotted with red, smaller than calyx lobes; pedicel ca. 1-1.5 cm long, appressed-pubescent. Calyx white, with red dots along margin, lobes free, erect, equal or subequal, ovate, 2.6 x 1.3 cm, margin repand-serrate, apex obtuse, outside and inside sparsely pubescent; corolla oblique in calyx, yellow with red markings inside, white outside, ca. 4 cm long, tube nearly cylindric at base, funnelform towards apex, 2.6 cm long, base spurred, 0.4-0.5 cm wide, middle widening towards throat, throat ventrally somewhat ventricose near limb, 1.8 cm wide, outside scattered pubescent, inside partially pubescent, limb ca. 2.5 cm wide, lobes subequal, spreading, suborbicular, 0.7-0.8 x 0.7-0.9 cm, margin subentire; stamens included, inserted at 0.3 cm from base of corolla tube; ovary ovoid, 0.6 x 0.4 cm, tomentose, style 2 cm long, hirto-pilose, stigma obscurely 2-lobed. Mature capsule ovoid, 1.5 x 0.9 cm, green to light brown.

Distribution: Southeastern Suriname and central French Guiana to northern Brazil (Pará); epiphytic on trees in rainforest, at 130-175 m alt.; 9 collections studied (SU: 5; FG: 3).

Selected specimens: Suriname: near the airstrip at the Oelemari R., Wessels Boer 989 (K, U, US), 1112 (U, US). French Guiana: Station de l' Arataye, Vieillescazes 513 (CAY, P); Takawana, Haut Oyapock, de Granville B-5236 (CAY, P, US).

Phenology: Collected in flower in March, April, and May.

2. **Drymonia coccinea** (Aubl.) Wiehler, Phytologia 27: 324. 1973. – *Besleria coccinea* Aubl., Hist. Pl. Guiane 2: 632, 4: pl. 255. 1775. – *Alloplectus coccineus* (Aubl.) Mart., Nov. Gen. 3: 189. 1832. – *Lophalix coccinea* (Aubl.) Raf., Sylva Tellur. 71. 1838. – *Columnea coccinea* (Aubl.) Kuntze, Revis. Gen. Pl. 2: 472. 1891. – *Crantzia coccinea* (Aubl.) Fritsch in Engl. & Prantl, Nat. Pflanzenfam. 4(3b): 168. 1894. Type: French Guiana, Cayenne, Aublet s.n. (holotype BM, isotype FI-W).

Alloplectus patrisii DC., Prodr. 7: 545. 1839. – *Macrochlamys patrisii* (DC.) Decne., Rev. Hort. (Paris) ser. 3. 3: 243. 1849. – *Columnea patrisii* (DC.) Kuntze, Revis. Gen. Pl. 2: 472. 1891. – *Crantzia patrisii* (DC.) Fritsch in Engl. & Prantl, Nat. Pflanzenfam. 4(3b): 168. 1894. Type: French Guiana, Cayenne, Patris s.n. (holotype G-DC).
Alloplectus coccineus (Aubl.) Mart. var. *fuscomaculatus* Leeuwenb., Acta Bot. Neerl. 7: 300, 361, fig. 8. 1958. Type: Suriname, Nassau Mts., Cowan & Lindeman 39049 (holotype U).

Terrestrial or epiphytic shrub, 0.6-2.5 m tall. Stem woody at base, succulent above, sarmentose, puberulous at apex, glabrous and shining. Leaves subequal or unequal in a pair; petiole (0.5-)1-4(-6) cm long, sparsely appressed-pubescent; blade chartaceous when dry, obliquely elliptic to oblong, variable, larger blade (2.7-)8-15(-29) x (1.8-)3-6(-10) cm, margin entire or sometimes obscurely denticulate near apex, apex acuminate, base cuneate to almost decurrent, above sparsely appressed-pubescent to glabrous, below sparsely appressed-pubescent to glabrous. Flowers solitary, or in short-racemose few-10 or more-flowered inflorescences; pedunculate, often with branched axis, each 0.2-6 cm long, sparsely appressed-pubescent; bracts numerous, large, pink, slightly cordate; pedicel 0.1-0.3 cm long, sparsely appressed-pubescent. Calyx colored like bracts, lobes free, subequal, leafy, 1.3-2.5 x 0.6-1.5 cm, margin entire, sinuate, or crenulate, apex obtuse or rounded, outside and inside sparsely appressed-pubescent, 4 erect, dorsal lobe recurved; corolla transversal-oblique in calyx, creamy, yellow or white, 3-4.8 cm long, tube nearly cylindric, 2.8-3.5 cm long, base spurred, 0.4-0.6 cm wide, middle straight to slightly curved, throat slightly contracted, 0.6-1.2 cm wide, outside puberulous to villous, sometimes glabrous at base, inside dorsally with an area of glandular hairs, limb 0.8-2.6 cm wide, lobes subequal, spreading, suborbicular, 0.4-1 x 0.4-1 cm, margin entire or sinuate; stamens included, inserted on corolla base; ovary ovoid, 0.4-0.6 x 0.3-0.5 cm, appressed-pubescent, style 3-3.5 cm long, appressed-pubescent, stigma capitate. Mature capsule hidden by bracts and calyx, yellow (acc. Aublet 1775, p. 632), subglobose, 1.3-1.5 x 0.9-1 cm.

Distribution: Colombia to Bolivia, eastern Venezuela to the Guianas, northern and central Brazil; usually epiphytic on trees in rainforests, at 0-700 m alt.; > 300 collections studied (GU: 20; SU: 40; FG: > 200).

Selected specimens: Guyana: Rupununi Distr., Kuyuwini Landing, Kuyuwini R., Jansen-Jacobs *et al.* 2316 (BBS, U, US); Rewa R., Clarke 3646 (U, US). Suriname: Maratakka R., Snake Cr., Maas & Tawjoeran LBB 10876 (BBS, U, US); Moengo, Jonker & Jonker 486 (U, US). French Guiana: Cr. Grégoire, Station Hydrologique, Deward 31 (CAY, US); Mt. Bellevue de l'Inini, de Granville *et al.* 7977 (B, CAY, INPA, MG, MO, P, U, US).

Phenology: Collected in flower during all months of the year, in fruit in April and August.

Vernacular names: French Guiana: crète poule (Créole); alalaka'a (Wayãpi); masakala kulumenay (Wayãpi; Jacquemin 1533); takaakabesu (Palikur; Grenand *et al.* 1987); waku djemba (Sar.; Sauvain 300); opaïpanga (Boni; Fleury 649).

Use: External febrifuge (Wayãpi).

Notes: Cultivated in several botanical gardens, and in limited cultivation among amateurs in North America.

Photographs: Feuillet & Skog, 2002 (pl. 65 d (unvouchered)); Grenand *et al.* 1987 (between pp. 256-257).

3. **Drymonia psilocalyx** Leeuwenb., Phytologia 48: 437. 1981. – *Drymonia psila* Leeuwenb., Misc. Pap. Landbouwhogeschool 19: 239, fig. 1. 1980 (non Gibson 1972). Type: French Guiana, Saül, de Granville 2000 (holotype WAG, isotypes CAY, P(2), WAG).
– Fig. 10 A-C

Epiphytic herb, 30-50 cm tall. Stem sappy, oblique, glabrous. Leaves strongly unequal in a pair; larger leaf in a pair: petiole 1-3 cm long, glabrous; blade papyraceous when dry, narrowly elliptic, 10-24 x 2.5-6.5 cm, margin entire, apex acuminate, base obliquely cuneate, glabrous on both faces; smaller leaf in a pair sessile, 0.5-5 x 0.3-1.5 cm, glabrous. Flowers solitary; epedunculate; pedicel 0.5-1.2 cm long, glabrous. Calyx pale green, lobes connate at base, 4 erect, dorsal lobe recurved, unequal with dorsal lobe much smaller, obliquely ovate, 2.2-2.7 x 0.8-1 cm, margin entire, apex obtuse, outside and inside glabrous; corolla oblique in calyx, creamy, 4 cm long, tube wide tubular, 2.7 cm long, base spurred, 0.5 cm wide, middle slightly curved, throat not contracted, 0.7 cm wide, outside glabrous, inside glabrous, limb 1.8 cm wide, lobes spreading, suborbicular, 0.5-0.8 x 0.5-0.8 cm, margin fimbriate, lower lobe larger, oblique; stamens included, inserted on corolla base; ovary obliquely ovoid, laterally compressed, 0.7 x 0.3 or 0.25 cm, apex minutely appressed-pubescent, style 0.8 cm long, glabrous, stigma large, capitate. Mature capsule dark violet, nearly black inside, obliquely ovoid, laterally compressed, 1.8 x 1 or 0.7 cm.

Distribution: Endemic to Suriname and French Guiana; epiphytic on trees in rainforests, at 200-780 m alt.; 36 collections studied (SU: 1; FG: 35).

Selected specimens: Suriname: Mts. Tumuc-Humac, Inselberg Talouakem, de Granville *et al.* 12147 (CAY); Waamahpan Cr., de Granville 958 (CAY). French Guiana: Région de l'Inini, Mt. Atachi Bacca, 5 km N du sommet principal, de Granville *et al.* 10573 (B, CAY, MO, NY, P, U, US); Mt. Bellevue de l'Inini, Feuillet 3750 (CAY, US).

Phenology: Collected in flower in January, April and May, in fruit in August.

50

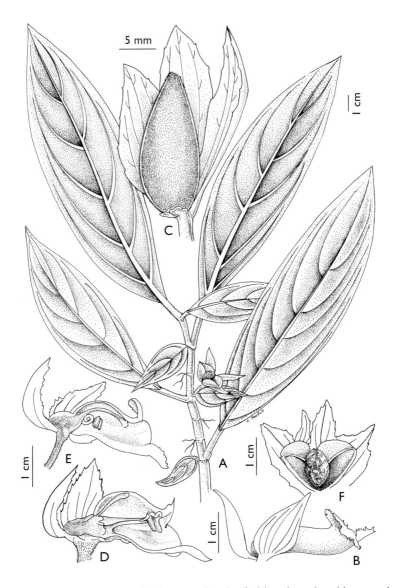

Fig. 10. *Drymonia psilocalyx* Leeuwenb.: A, fruiting branch with strongly unequal leaves; B, flower; C, calyx with 2 sepals removed to show nectary and young fruit. *Drymonia serrulata* (Jacq.) Mart.: D, flower with part of calyx and corolla removed to show mature stamens and young style; E, flower with part of calyx and corolla removed to show recoiled old stamens and mature style; F, fruit open in persistent calyx. (A, de Granville 1615; B, Feuillet 3750; C, de Granville 1662; D, Feuillet 4659; E, from photo of cultivated plant; F, from photo of Feuillet 824).

4. **Drymonia serrulata** (Jacq.) Mart., Nov. Gen. Sp. Pl. 3: 192. 1832.
– *Besleria serrulata* Jacq., Pl. Hort. Schoenbr. 3: 21, pl. 290. 1798.
Type: Cult. Hort. Schoenbr., Jacquin s.n., sterile (holotype W).
– Fig. 10 D-F

Drymonia calcarata Mart., Nov. Gen. Sp. Pl. 3: 58, pl. 224. 1829. – *Besleria drymonia* Steud., Nomencl. Bot. ed. 2. 1: 200. 1840, (non *Besleria calcarata* Kunth 1818). Type: Brazil, Amazonas, Martius 3014 (lectotype M) (designated by Leeuwenberg 1958: 308).
Drymonia cristata Miq., Linnaea 18: 26. 1844. Type: Suriname, Focke 766 (holotype U, flowers lost).

Terrestrial or epiphytic herb, low shrub, or more usually a liana, when erect to 2 m tall, or up to 10 m long. Stem succulent when young, becoming subwoody, erect or scandent, sparsely strigose or puberulous towards apex. Leaves equal to nearly equal in a pair; petiole 0.6-3 cm long, appressed pubescent; blade chartaceous when dry, elliptic or oblong to obovate, 4-19 x 2-6 cm, margin serrulate to sinuate-dentate, apex acute or acuminate, base cuneate to rounded, sometimes oblique, above sparsely strigillose to scabrous, below sparsely strigillose to glabrescent, less between veins. Flowers solitary or rarely 3 in a fascicle; epedunculate; pedicel 0.5-2.5 cm long, puberulous. Calyx green, lobes more or less free, 4 erect, subequal, leaf-like, ovate, lanceolate to oblong, 2-5 x 0.5-3.0 cm, margin all entire, sinuate to serrulate, apex acute or acuminate, dorsal lobe recurved and smaller, outside and inside puberulous or sparsely strigillose; corolla oblique in calyx, variously colored, from white to pale yellow or dark red, 3-7 cm long, tube wide tubular, 2-3 cm long, base spurred, 0.3-0.8 cm wide, middle gibbous dorsally, broadening towards throat, throat round, 1-2 cm wide, outside puberulent, inside glandular-pubescent at least towards throat, limb 3-4 cm wide, lobes spreading, rounded, 1-1.5 x 1-1.5 cm, margin serrulate, except lower lobe erose, larger and oblique; stamens included, inserted on corolla base; ovary ovoid, 0.5-0.8 x 0.4-0.6 cm, puberulous, style ca. 2.5 cm long, glandular-pubescent, stigma 2-lobed. Mature capsule yellowish to purplish outside, orange to red inside, globose to broadly ovoid, 1-2 x 1-2 cm.

Distribution: Widely distributed from Mexico to Matto Grosso in Brazil, and in the Lesser Antilles; sometimes an epiphyte, but often rooted in the ground and climbing on trees or shrubs, over rocks on exposed outcrops, or in moist undisturbed forests, and even at roadsides, clearings, and in second growth forests, at 40-700 m alt.; > 80 collections studied (GU: 1; SU: 15; FG: 20).

Selected specimens: Guyana: Mabura Region, W Pibiri, Ek *et al.* 942 (U, US). Suriname: Lely Mts., SW plateaus, along airstrip, Lindeman & Stoffers *et al.* 447 (BBS, U, US); Saramaca R., above Kwatta Hede, Maguire 23935 (NY, US). French Guiana: Grand Santi, Daniel 1 (CAY, US); Région de Saül, Savane-roche Dachine, Cremers *et al.* 14645 (CAY, NY, P, U, US).

Phenology: Collected in flower and fruit during most months of the year.

Vernacular name: French Guiana: opoidjemba (Ndjuka; Sauvain 424).

10. **EPISCIA** Mart., Nov. Gen. Sp. Pl. 3: 39. 1829.
Type: E. reptans Mart.

Cyrtodeira Hanst., Linnaea 26: 207. 1854 ('1853').
Type: C. cupreata (Hook.) Hanst. (Achimenes cupreata Hook.)
Episcia Mart. subsect. *Tremanthera* Leeuwenb., Acta Bot. Neerl. 7: 309. 1958.
Type: E. sphalera Leeuwenb.

Terrestrial or epiphytic, caulescent, decumbent, creeping, or sprawling low herbs, rarely subshrubs, stoloniferous. Stems often branched. Leaves opposite, often crowded, usually nearly equal in a pair, venation pinnate, foliar nectaries absent. Flowers axillary, 1-6 in cymose inflorescences; pedunculate or epedunculate; bracteoles present; pedicellate. Calyx lobes free or briefly connate at base, 4 lobes erect, the 5th dorsal, curved around corolla spur; corolla white, yellow, blue, purple to red, tubular, salver-shaped to campanulate; stamens included, filaments basally connate, anthers coherent in pairs in a square or arc, becoming free, dehiscing by longitudinal slits, thecae parallel or divergent; staminode minute; disc a single large dorsal gland; ovary superior, stigma stomatomorphic, 2-lobed, or capitate. Fruit a fleshy, green or brown to reddish brown capsule, loculicidally dehiscent, 2-valved, valves opening widely.
Chromosome number n=9 (Skog 1984).

Distribution: A genus of 10-15 species in C America and northern S America; usually in moist forest, in well drained, light shade situations; 3 species in the Guianas.

Note: Included in the key are 2 species (*E. cupreata* (Hook.) Hanst. and *E. lilacina* Hanst.) which are cultivated and possibly escaping.

KEY TO THE SPECIES

1 Corolla red 2
   Corolla yellow, lavender, or white 3

2 Corolla tube 3-3.5 cm long, nearly straight, roseate in the throat, 5 lobes
   spreading . . . . *1. E. reptans*

   Corolla tube 2.5 cm long, curved at middle, yellow in the throat, 2 dorsal lobes
   recurved (cultivated) . . . . *E. cupreata*

3 Corolla yellow; calyx lobes obovate . . . . *3. E. xantha*
   Corolla white or lavender; calyx lobes lanceolate or spathulate 4

4 Leaf blade pale green; corolla white; calyx lobes lanceolate . . . *2. E. sphalera*
   Leaf blade dark with pale markings along the main veins; corolla tube white
   with lavender lobes; calyx lobes spathulate (cultivated) . . . . . . *E. lilacina*

1. **Episcia reptans** Mart., Nov. Gen. 3: 41, pl. 217. 1829. Type: Brazil, Amazonas, Japura, Martius 3091 (lectotype M) (designated by Leeuwenberg 1958: 412).

Terrestrial or epiphytic herb, about 5-20 cm tall, to 50 cm long. Stem sappy, creeping, hirsute. Leaves equal or subequal in a pair; petiole 0.5-7 cm long, hirsute; blade chartaceous when dry, elliptic, 2-13(-15) x 1.3-9(-10) cm, margin crenate-serrate, apex acute or obtuse, base rounded or subcordate, above and below hirsute. Flowers 1-3 in fascicles; peduncle up to 0.5 cm long, hirsute; pedicel 1-4 cm long, hirsute. Calyx green, lobes nearly free, subequal, lanceolate, slightly narrowed towards base, (0.7-)0.9-1.3(-1.5) x 0.2-0.4 cm, margin serrate near apex, apex acute, outside and inside hirsute; corolla oblique in calyx, scarlet, 3.5-4 cm long, tube trumpet-shaped, 3-3.5 cm long, base spurred, 0.3-0.4 cm wide, middle slightly and gradually curved, not ventricose, throat not contracted, slightly widened, 0.5-0.8 cm wide, outside hirsute, inside with a ring of glandular hairs in throat, limb 1.5-2.5 cm wide, lobes subequal, spreading, ventral lobe oblique, suborbicular, 0.5-1 x 0.5-1 cm, margin serrulate; stamens included, inserted on base of corolla tube; ovary ovoid, 0.4-0.5 x 0.25-0.3 cm, hirsute, style ca. 1 cm long, glabrous, stigma capitate. Mature capsule globose, 1 x 1 cm.

Distribution: Colombia, Peru, Venezuela, Guyana, and Brazil (Amazonas, Rio Branco, and Minas Gerais); on rocks or on tree trunks, among mosses, in rainforests, near waterfalls, at low elevations; 90 collections studied (GU: 3).

Specimens studied: Guyana: Between Demerara R. and Berbice R., de la Cruz 1588 (CM, K, US); Upper Mazaruni R., de la Cruz 2350 (CM, MO, US); Rupununi Distr., near Dadanawa, de la Cruz 1787 (CM, MO, US).

Phenology: Collected in flower in July, September, and October.

2. **Episcia sphalera** Leeuwenb., Acta Bot. Neerl. 7: 310, 413, fig. 27. 1958. Type: Suriname, Kappler 2044 (holotype P, isotypes GOET, RO, W(2)).

Terrestrial herb, less than 10 cm. Stem sappy, creeping, hirsute. Leaves equal or subequal in a pair; petiole 0.5-4 cm long, hirsute; blade membranaceous when dry, elliptic or oblong-ovate, 2-7 x 1.5-4 cm, margin crenate-serrate, apex acute or obtuse, base rounded or subcordate, above and below hirsute. Flowers solitary; epedunculate; pedicel 2-5 cm long, hirsute. Calyx green, lobes nearly free, lanceolate, slightly narrowed towards base, 0.6-0.9 x 0.15-0.3 cm, margin entire or with some teeth, apex acute, outside hirsute, inside hirsute, 4 lobes subequal, dorsal lobe smaller; corolla oblique in calyx, white with purple lines in tube, 2.4-2.5 cm long, tube obliquely infundibuliform, 2-2.5 cm long, base spurred, 0.3 cm wide, middle curved, not ventricose, throat not or hardly contracted, 0.6-1 cm wide, outside villous, inside with a ring of glandular hairs in throat, limb 2-2.5 cm wide, lobes subequal, spreading, ventral lobe oblique, suborbicular, 0.6-1 x 0.6-1 cm, margin crenate-serrate; stamens slightly exserted, inserted on base of corolla tube; ovary ovoid, 0.25 x 0.15 cm, hirsute, style ca. 1 cm long, pubescent with glandular hairs, stigma saucer-shaped or obscurely 2-lobed. Mature capsule globose, 0.5 x 0.5 cm.

Distribution: Suriname, French Guiana, and northern Brazil (Amapá, Amazonas, Pará); terrestrial, on white sand, river banks, on rocks, or lower part of tree trunks, at 100-450 m alt.; 21 collections studied (SU: 1; FG: 12).

Selected specimens: Suriname: no locality, Kappler 2044 (GOET, P, RO, W(2)). French Guiana: Lower Mana R., near Godebert, Wachenheim 209 (P); Massif des Emérillons, de Granville 3784 (CAY, P); Saut Grand Canori, Oldeman 2757 (CAY, P, US).

Phenology: Collected in flower in January, February, May, July-September, and December.

3. **Episcia xantha** Leeuwenb., Misc. Pap. Landbouwhogeschool 19: 241, fig. 2. 1980. Type: French Guiana, Mts. de Kaw, Leeuwenberg 11819 (holotype WAG, isotypes CAY, K, MO, P, U, US).

– Fig. 11

Terrestrial herb, less than 10 cm. Stem sappy, decumbent, hirto-pilose. Leaves equal or subequal in a pair; petiole 1-8 cm long, hirsute; blade membranaceous when dry, elliptic or ovate, 5-20 x 3-15 cm, margin crenate-serrate, apex obtuse or acute, base rounded, subcordate, or less often cuneate, above hirto-pilose, below sparsely hirto-pubescent. Flowers few to many in cymes; peduncle 1.5-5 cm long, sparsely pilose or glabrous; pedicel 0.3-1.2 cm long, pilose or glabrous. Calyx pale green, lobes free, obovate, gradually narrowed towards base, 1-1.2 x 0.5-0.8 cm, margin conspicuously ciliate, apex obtuse, with 2-5 teeth, outside and inside sparsely and minutely pubescent, dorsal lobe narrower than other 4; corolla oblique in calyx, yellow, 2.6 cm long, tube obliquely infundibuliform, 1.6 cm long, base spurred, 0.3 cm wide, middle curved and becoming wider, throat not contracted, 0.7-0.8 cm wide, outside hirsute in upper half, glabrous in lower, inside partially pubescent with glandular hairs, limb 2 cm wide, lobes subequal, spreading, broadly suborbicular, 0.5 x 0.6-0.7 cm, margin entire; stamens included, inserted on base of corolla tube; ovary ovoid, laterally compressed, 0.3-0.4 x 0.2-0.3 or 0.15-0.25 cm, hirsute except for glabrous base, style 1 cm long, glabrous, stigma large, capitate. Mature capsule light brown, subglobose, laterally compressed, 1 x 0.8-0.9 or 0.6 cm.

Distribution: Endemic to the Guianas (Guyana, French Guiana); on rotting logs or terrestrial in forests, at 50-550 m alt.; 63 collections studied (GU: 3; FG: 60).

Selected specimens: Guyana: Potaro-Siparuni Region, Iwokrama Rainforest Reserve, Mori *et al.* 24647 (NY), Clarke *et al.* 4260 (U, US). French Guiana: Mt. Atachi Bacca, de Granville *et al.* 10638 (CAY, B, US); Mt. Bellevue de l'Inini, Feuillet 3699 (CAY, US).

Phenology: Collected in flower and fruit from January to September.

Vernacular name: French Guiana: tapu (Wayana; de Granville *et al.* 7851).

Notes: Seeds often germinate in the capsule.
Photograph: Feuillet & Skog, 2002 (pl. 65 b (Mori *et al.* 22285)).

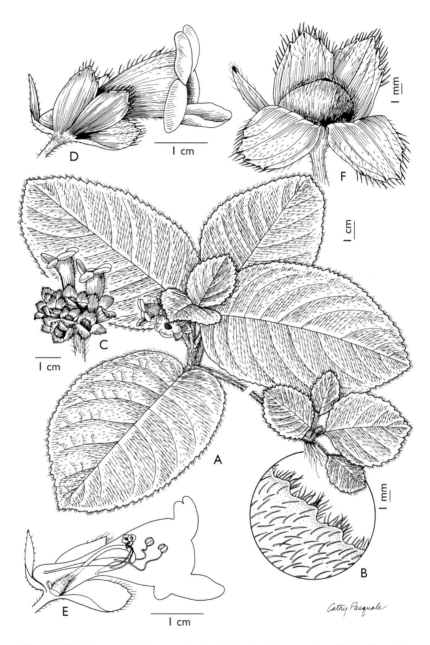

Fig. 11. *Episcia xantha* Leeuwenb.: A, habit showing a stolon; B, leaf margin; C, inflorescence; D, flower; E, calyx and corolla with parts removed to show stamens, nectary, and pistil; F, young fruit in calyx. (A-F, Feuillet *et al.* 10118).

11. **GLOXINIA** L'Hér. in Aiton, Hort. Kew. 2: 331. 1789.
Type: G. maculata L'Hér., nom. illeg. (Martynia perennis L., Gloxinia perennis (L.) Fritsch)

Seemannia Regel, Gartenflora 4: 183. 1855, nom. cons.
Type: S. ternifolia Regel [Gloxinia sylvatica (Kunth) Wiehler]

Terrestrial, caulescent, decumbent to erect herbs, with scaly rhizomes. Stems seldom branched. Leaves opposite, subequal to equal in a pair, venation pinnate, foliar nectaries absent. Flowers axillary, 1 to several in fascicles; epedunculate; bracteoles absent (in Guianan species); pedicellate. Calyx lobes nearly free; corolla white, blue, lavender, purple, rose-pink, scarlet to orange or yellow, often with darker spots, tubular, cylindric to campanulate; stamens included, filaments not connate, anthers coherent by their apices, dehiscing by longitudinal slits, thecae parallel or divergent at base; staminode present; disc absent, or when present, annular, entire or 5-lobed; ovary inferior, stigma stomatomorphic or 2-lobed. Fruit a dry, brown capsule, loculicidally dehiscent, 2-valved, valves opening slightly.
Chromosome number n=13 (Skog 1984).

Distribution: About 10 species distributed in the rainforests of C and S America, from Panama to Bolivia and Argentina, and to the Guianas and Trinidad; 2 species in the Guianas.

Notes: In tropical regions, some species in cultivation as ornamentals occasionally escape and may become weedy.
The plants in cultivation as the 'Florist's Gloxinia' are cultivars of *Sinningia speciosa* (Lodd.) Hiern.

LITERATURE

Hoehne, F.C. 1964. O genero Gloxinia no Brasil. Arq. Bot. Estado São Paulo 3: 315-335.
Roalson, E.H., J.K. Boggan, L.E. Skog & E.A. Zimmer. 2005. Untangling Gloxinieae (Gesneriaceae). Phylogenetic patterns and generic boundaries inferred from nuclear, chloroplast, and morphological cladistic datasets. Taxon 54: 389-410.
Wiehler, H. 1976. A report on the classification of Achimenes, Eucodonia, Gloxinia, Goyazia, and Anetanthus (Gesneriaceae). Selbyana 1: 374-404.

KEY TO THE SPECIES

1 Leaves opposite, petiole 1.4-12 cm long, blade truncate, rounded to cordate
at base; calyx lobes ovate to oblong, seldom lanceolate; corolla broadly
campanulate .... *1. G. perennis*
Leaves in whorls (only rarely opposite), petiole usually 1 cm long, blade
acute or cuneate at base; calyx lobes lanceolate; corollas cylindric to
ventricose .... *2. G. purpurascens*

1. **Gloxinia perennis** (L.) Fritsch in Engl. & Prantl, Nat. Pflanzenfam.
4(3b): 174. 1894. – *Martynia perennis* L., Sp. Pl. 618. 1753. Type:
Cult. Hort. Cliffortianus (holotype BM).

Gloxinia suaveolens Decne., Rev. Hort. (Paris) ser. 3. 2: 463. 1848. – *Salisia
suaveolens* (Decne.) Regel, Bot. Zeitung (Berlin) 9: 894. 1851. Type:
"Guyane" (not seen).
Gloxinia trichantha Miq., Linnaea 22: 473. 1849. Type: Suriname,
Paramaribo, Focke 822 (holotype U, isotype U).

Terrestrial herb, to 1 m tall. Stem subwoody at base, succulent above,
erect or ascending, nearly glabrous. Leaves subequal in a pair; petiole
1.4-12.0 cm long, sparsely pilose; blade papyraceous when dry,
orbicular, ovate to rarely obovate, 5.5-14.7(-18) cm x 3.4-9.5(-14) cm,
margin coarsely crenate to serrate, apex obtuse to acute, base sometimes
oblique, truncate, rounded to usually cordate, above glabrous to sparsely
strigose, below glabrous or with a few scattered hairs. Flowers appearing
terminal, but actually solitary on a raceme-like stem; epedunculate;
pedicel 0.5-4.0 cm long, more or less glabrous. Calyx spreading
campanulate, light green, sometimes streaked with red, lobes free,
spreading, subequal, lanceolate to oblong, 0.7-1.9 x 0.4-0.8 cm, margin
entire or toothed, apex broadly acute, outside and inside glabrous;
corolla oblique in calyx, white, pink, lavender to purple, 2.5-4.0 cm long,
tube broadly campanulate, 2-3.5 cm long, base broad, ventrally gibbous,
ca. 1 cm wide, middle broadly ventricose, throat slightly contracted, 2-
3.2 cm wide, outside pilose, inside glandular on upper surface, limb to
4 cm wide, lobes subequal, upper and lateral lobes spreading, basal lobe
incurved, all broadly suborbicular, 1.0-1.2 x 1.0-1.8 cm, margin upper
and lateral lobes entire, basal lobe toothed; stamens included, inserted at
base of corolla tube; ovary apex broadly ovoid, 0.3-0.5 x 0.3-0.4 cm,
apex pubescent, style 1.2-1.5 cm long, glandular, sparsely pilose, stigma
stomatomorphic. Mature capsule green to brown, narrowly conic, 1-1.8
x 0.4-0.7 cm.

Distribution: Panama south to Bolivia and Brazil, east to Trinidad and the Guianas; on wet rocks in rainforest; < 200 specimens studied, with only 10 in the Guianas (GU: 1; SU: 4; FG: 5).

Selected specimens: Guyana: Cultivated in Botanic Gardens, Georgetown, collector unknown s.n. (BRG). Suriname: "Horto culta et arenosis spontanea", Paramaribo, Focke 822 (U(2)); Cult., Paramaribo, Wullschlägel 767 (BR, U), s.n. (BR), s.n. (BR). French Guiana: St. Georges de l'Oyapock, Benoist s.n. (P); Leprieur, ann. 1850 (P(2)); St. Georges, Lemée s.n. (P); Richard s.n. (P); Cult., Poiteau s.n. (LE).

Phenology: Unknown.

Note: The wild-collected Guianan specimens may have escaped from cultivation and become naturalized. The species has been in cultivation for over 270 years in tropical gardens and as pot-plants in temperate regions.

2. **Gloxinia purpurascens** (Rusby) Wiehler, Selbyana 1: 387. 1976. – *Seemannia purpurascens* Rusby, Mem. Torrey Bot. Club 4: 237. 1895. – *Fritschiantha purpurascens* (Rusby) Kuntze, Revis. Gen. Pl. 3(3): 241. 1898. Type: Bolivia, Yungas, Bang 542 (holotype NY, isotypes BM, E, F, GH, K, MANCH, MO, PH, US, W). – Fig. 12

Terrestrial herb, to 50 cm or more tall. Stem subwoody at base, succulent above, erect or somewhat weak but ascending, pilose. Leaves subequal in a pair or whorl; petiole usually less than 1cm long, strigillose; blade papyraceous when dry, lanceolate to ovate or elliptic, 4-7 x 2-9 cm, margin finely serrulate, ciliate, apex acute to acuminate, base acute to cuneate, above sparsely pilose to subscabrous, below strigillose along obvious veins. Flowers solitary on a raceme-like stem; epedunculate; pedicel 2-4(-7) cm long, strigillose. Calyx green, lobes free, erect to spreading, subequal, lanceolate, 0.7-1 x 0.2-0.3 cm, margin entire, apex long-acuminate, outside pubescent, inside pubescent; corolla slightly oblique in calyx, orange to rose-purple, ca. 3 cm long, tube cylindric to ventricose, 2.5-3.3 cm long, base not spurred, 0.4-0.6 cm wide, middle ventricose above, throat contracted at mouth, 0.7-0.8 cm wide, outside densely pilose or villous, inside sparsely glandular pubescent, limb ca. 1.2 cm wide, lobes subequal, spreading, broadly triangular, ca. 0.1 cm long, wide, margin entire; stamens subincluded, inserted at 0.1-0.2 cm from base of corolla; ovary ovoid-conic, 0.2-0.3 x 0.3-0.4 cm, appressed-pubescent, style 2-3 cm long, pubescent, stigma ovoid. Mature capsule brown, obconic, ca. 0.5 x ca. 0.5 cm.

60

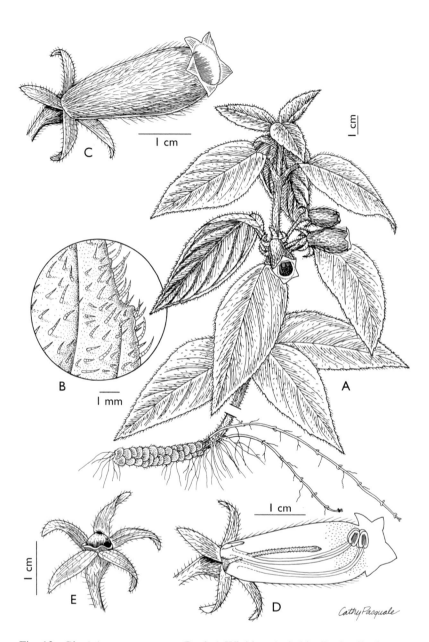

Fig. 12. *Gloxinia purpurascens* (Rusby) Wiehler: A, habit; B, detail of upper leaf surface; C, flower; D, flower with part of corolla removed to show stamens and pistil; E, young fruit surrounded by persistent calyx and nectary. (A-D, Boom *et al*. 8843; E, Kvist 245).

Distribution: Bolivia and Peru, and the Guianas to northern Brazil; disturbed forest and savannas, at 450-670 m alt.; > 40 specimens studied, with 3 in the Guianas (GU: 2; FG: 1).

Specimens studied: Guyana: Potaro-Siparuni Region: near Kopinang village, Boom & Samuels 8843 (US); Chinapou, 50 km upstream from Kaieteur Falls, Kvist et al. 245 (AAU, B, BBS, BRG, CAY, COL, K, NY, P, U, US). French Guiana: sur le Tampoc, Saut Koumarou, Cremers 4514 (CAY, US).

Phenology: Collected in flower in March, April, June, and October.

Note: This species is quite variable throughout its range in habit, size, and corolla color.

12. **KOHLERIA** Regel, Ind. Sem. Hort. Bot. Turic. [4]. 1847.
    Type: K. hirsuta (Kunth) Regel (Gesneria hirsuta Kunth)

Terrestrial or epipetric, caulescent, decumbent to erect herbs, subshrubs or shrubs, with scaly rhizomes. Stems rarely branched. Leaves opposite, rarely ternate, equal to subequal in a pair, venation pinnate, foliar nectaries absent. Flowers 1-6(-10) in axillary fasciculate or cymose inflorescences; epedunculate or rarely pedunculate; bracteoles small, but often caducous; pedicellate. Calyx lobes connate for ca. $^1/_3$ their length, rarely free; corolla orange-red to orange-yellow, with red spots on limb or throat, funnelform or cylindric; stamens included to subincluded, filaments not connate, anthers coherent at apices and sides, dehiscing by longitudinal slits, thecae parallel; staminode present; disc usually of 5 free glands or of 3 free and 2 basally united glands or a 5-lobed ring; ovary half-inferior to nearly completely inferior, stigma 2-lobed. Fruit a dry, brown capsule, loculicidally dehiscent, 2-valved, valves opening slightly. Chromosome number n=13 (Skog 1984).

Distribution: A genus of 17 species ranging from Mexico to Peru and east to Suriname, in wet forests and forest edges; 1 species in the Guianas.

Note: In the Botanic Garden in Georgetown also K. tubiflora (Cav.) Hanst. has been cultivated, originally from Costa Rica, Panama, Colombia, and Venezuela, with a red cylindric corolla.

Uses: Species and cultivars of Kohleria are commonly cultivated in temperate parts of the world as ornamentals. Plant parts are used in folk medicine in Andean countries (Kvist 1986).

LITERATURE

Kvist, L.P. and L.E. Skog. 1992. Revision of Kohleria (Gesneriaceae). Smithsonian Contr. Bot. 79: 1-83.

1. **Kohleria hirsuta** (Kunth) Regel, Flora 31: 250. 1848. – *Gesneria hirsuta* Kunth in Humboldt, Bonpland & Kunth, Nov. Gen. Sp. ed. fol. 2: 317, pl. 189. 1818 ('1817'). – *Isoloma hirsutum* (Kunth) Regel, Bot. Zeit. (Berlin) 9: 893. 1851. – *Brachyloma hirsutum* (Kunth) Hanst., Linnaea 26: 203. 1854 ('1853'). Type: Venezuela, Monagas, Bonpland 330 (holotype P-B, isotypes B-W, P).

In the Guianas only: var. **hirsuta**   – Fig. 13

Terrestrial (frequently epipetric) herb or subshrub, 30-150 (-200) cm tall. Stem subwoody at base, succulent above, erect or ascending, hirsute to villous above, glabrescent below. Leaves equal or subequal in a pair; petiole 1-4(-8) cm long, lanate-hirsute; blade papyraceous when dry, elliptic, less commonly ovate to elliptic, 4-12 x 2-6(-10) cm, margin serrate, apex acuminate, base acute to subcordate, above pubescent to sericeous, below villous to tomentose, veins sericeous or villous. Flowers 1 to 4(-6) in fascicles or cymes; usually epedunculate or less commonly pedunculate, peduncle, when present, 0-2(-4) cm long, lanate-hirsute; pedicel 2-6(-10) cm long, lanate-hirsute. Calyx campanulate, green, rarely purplish, lobes free, erect, subequal, lanceolate to subulate, rarely triangular, 0.6-1.3 x 0.1-0.6 cm, margin entire, apex acute to acuminate, outside villous or sericeous, inside appressed pubescent to sericeous; corolla erect in calyx, orange-red, 2.2-5 cm long, tube funnelform, subventricose, (1.8-)2.5-3.8(-4.5) cm long, base usually saccate but not sharply delimited, 0.2-0.6(-1) cm wide, middle subventricose, throat slightly contracted, 0.3-1.5 cm wide, outside villous, inside glabrous, limb 0.6-2.2 cm wide, lobes subequal, spreading, suborbicular, 0.3-0.6 cm long, at base 0.4-0.7 cm wide, margin entire; stamens subincluded, inserted at base of corolla tube; ovary ovoid-conic, 0.4-0.7 x 0.3-0.6 cm, hirsute, style (2.2-)2.7-3.3 (-3.7) cm long, pubescent, stigma 2-lobed. Mature capsule brown, ovoid, 1-1.4 x 0.5-0.9 cm.

Distribution: Ecuador, Colombia, Venezuela, Trinidad, and the Guianas; often on exposed, dry granitic rocky slopes, or wet sandstone cliffs, at 400-1100 m alt.; > 250 collections seen, 12 from the Guianas (GU: 11; SU: 1).

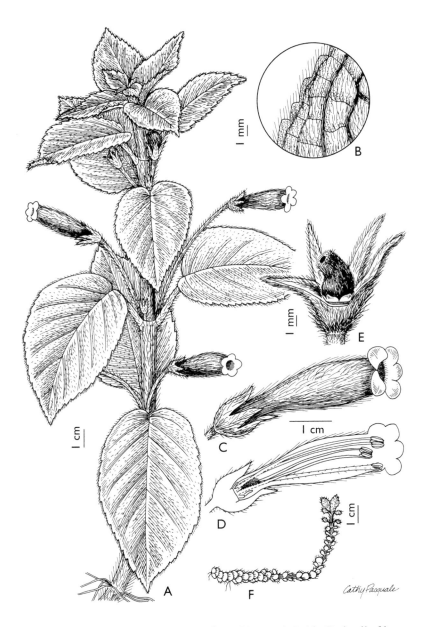

Fig. 13. *Kohleria hirsuta* (Kunth) Regel var. *hirsuta*: A, habit; B, detail of lower leaf surface and margin; C, flower; D, flower with corolla opened to show stamens and pistil; E, young fruit with nectary at base; F, rhizomes with shoot at apex. (A, Tillett *et al.* 45809; B, Tillett *et al.* 44974; C-D, Hahn *et al.* 4658; E, Goodland *et al.* 456A).

Selected specimens: Guyana: Cuyuni-Mazaruni Region: Upper Mazaruni R. basin, Kamarang R., Tillett *et al.* 45809 (NY, US); Mt. Ayanganna, Tillett *et al.* 44974 (NY, US); Potaro-Siparuni Region: Kaieteur National Park, Gillespie *et al.* 1293 (US); Upper Takutu-Upper Essequibo Region: Kanuku Mts., Iraimakipang Summit, Goodland & Maycock 456A (NY, US); same locality, Wilson-Browne 623 (BRG, K, NY). Suriname: Wilhelmina Mts., Julianatop, Schulz LBB 10315 (U).

Phenology: Collected in flower April-August and October, in fruit February and July.

Note: In the Guianas this species is represented by the typical and more widespread variety. *Kohleria hirsuta* var. *longipes* (Benth.) L.P. Kvist & L.E. Skog is restricted to Colombia.

13. **LAMPADARIA** Feuillet & L.E. Skog, Brittonia 54: 344. 2003 ('2002').
Type: L. rupestris Feuillet & L.E. Skog

Terrestrial, caulescent, erect or decumbent, small herbs, without modified stems. Stems unbranched. Leaves opposite, equal or subequal in a pair, venation pinnate, foliar nectaries absent. Flowers axillary, many in cymose inflorescences, long pedunculate, bracteoles leaf-like, pedicellate. Calyx lobes free; corolla white, campanulate; stamens included, filaments not connate or basally connate, anthers free, dehiscing by longitudinal slits, thecae parallel; staminode present but minute; disc of 2 opposite glands; ovary superior, stigma capitate. Fruit a fleshy, becoming dry, (color unknown) capsule, loculicidally dehiscent, 2-valved, valves opening slightly.
Chromosome number unknown.

Distribution: Monotypic, 1 species endemic in Guyana, in rain forest, on sandstone boulders.

1. **Lampadaria rupestris** Feuillet & L.E. Skog, Brittonia 54: 345. 2003 ('2002'). Type: Guyana, Potaro-Siparuni Region: near North Fork R., McDowell 4872 (holotype BRG, isotypes K, NY, U, US).
– Fig. 14

Terrestrial herb, to 15 cm tall (without inflorescence). Stem sappy, decumbent, tomentose. Leaves subequal in a pair; petiole 3-7.5 cm long, tomentose; blade chartaceous when dry, ovate-elliptic, 5.5-8 x 2.4-6 cm,

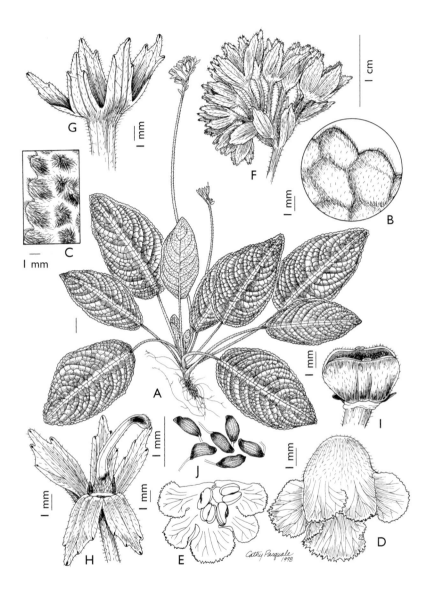

Fig. 14. *Lampadaria rupestris* Feuillet & L.E. Skog: A, habit; B, leaf margin from below; C, leaf margin from above; D, corolla (view from above); E, corolla opened dorsally to show stamens and staminode; F, inflorescence; G, calyx (side view); H, calyx, with two sepals bent to show nectary and pistil; I, capsule (calyx removed); J, seeds. (A-J, McDowell 4872).

margin crenate, apex barely acute to blunt, base rounded to slightly cordate, sometimes asymmetrical, above pilose, below tomentose on veins, loosely pilose between veins. Flowers many, in long pedunculate cymes; peduncle 15-20 cm long, with small, loose hairs; pedicel 0.6 cm long, pilose. Calyx lobes free, oblique, subequal, lanceolate, 0.4-0.5 x 0.2 cm, margin 3-5-toothed in distal ⅓, apex acute, outside and inside pilose; corolla erect in calyx, white, 0.7 cm long, tube campanulate, 0.3-0.4 cm long, base not spurred, 0.2 cm wide, middle slightly ventricose, not constricted, throat not constricted, 0.6-0.7 cm wide, outside shortly tomentose, inside glabrous, limb 1-1.5 cm wide, lobes subequal, spreading, ventral lobe less so, suborbicular, 0.2-0.35 x 0.2-0.35 cm, margin dentate-fimbriate; stamens included, inserted on corolla tube; ovary ovoid, 0.1 x 0.15 cm, hirsute, style 0.35-0.4 cm long, glabrescent, stigma capitate. Mature capsule light brown, laterally compressed, wider than long, < 0.3 x 0.5 cm.

Distribution: Endemic to central Guyana, near Mt. Ebini and Mt. Wokomung; rainforest, on sandstone boulders, at ca. 300 m alt.; known from only 2 collections.

Specimens examined: Guyana: Potaro-Siparuni Region, Upper Potaro R., Clarke 8897 (BRG, CAY, K, MO, NY, P, US); near North Fork R., McDowell 4872 (BRG, K, NY, U, US).

Phenology: Collected in flower and fruit in May.

14. **LEMBOCARPUS** Leeuwenb., Acta Bot. Neerl. 7: 318, 418. 1958.
Type: L. amoenus Leeuwenb.

Terrestrial or sometimes epiphytic, acaulescent, erect, small herbs, tuberous. Stems unbranched. Leaves opposite, radicular, strongly unequal in a pair, rarely more than one pair in the wild, plant appearing as having a single leaf, foliar nectaries absent; smaller leaf mostly scale-like, very small to minute, rarely a much smaller version of the larger one. Flowers axillary, few in subcymose inflorescences; pedunclulate; bracteoles small, pedicellate. Calyx lobes free; corolla pale blue or tube white and limb lavender or white with purple dots inside, broadly tubular or campanulate; stamens included, filaments not connate, anthers coherent in pairs, dehiscing by longitudinal slits, thecae parallel; staminode present; disc absent; ovary superior, stigma 2-lobed. Fruit a fleshy, becoming dry, green to light brown capsule, loculicidally dehiscent, 2-valved, valves opening to 180°.
Chromosome number unknown.

Distribution: Monotypic, 1 species from Suriname, French Guiana and northern Brazil; in rain forest on lateritic or granitic boulders.

1. **Lembocarpus amoenus** Leeuwenb., Acta Bot. Neerl. 7: 319, 418, fig. 28. 1958. Type: Suriname, Maguire 40806 (holotype U, isotypes NY, S).   – Fig. 15 A-D

Terrestrial or sometimes epiphytic herb, very small to ca. 10 cm tall (without inflorescence). Stem absent or very short. Leaves with the petiole 0.5-10 cm long, pilose or villous; blade thinly membranaceous when dry, cordate-orbicular, (1-)5-8(-14) x (0.7-)4.5-7.5(-14) cm, margin sinuate-dentate, apex acute or obtuse, base cordate to slightly oblique, above and below sparsely strigillose, especially on midrib and veins. Flowers 1-7 in pedunculate cymes; peduncle 4-16 cm long, pilose or villous; pedicel 0.5-2 cm long, pilose or villous. Calyx green, lobes free, spreading, equal, narrow-triangular, 0.5-1.1 x 0.05-0.3 cm, margin entire or obscurely sinuate-dentate, apex acuminate, outside villous with ordinary and capitate hairs, inside minutely puberulous or glabrous; corolla erect in calyx, pale blue or tube white and limb lavender, 1.6-2.5 cm long, tube campanulate, 1.2-2 cm long, base not spurred, 0.2-0.4 cm wide, middle not curved or contracted, throat not contracted, 0.9-1.6 cm wide, outside pilose or villous, inside glabrous(?), limb 1.1-2.1 cm wide, lobes subequal, oblique, broadly rounded, 0.5-1.2 x 0.2-0.6 cm, margin entire, ciliolate; stamens included, inserted at base of corolla tube; ovary ovoid, 0.2-0.25 x 0.15-0.2 cm, puberulous, style 0.5-0.7 cm long, glabrous, stigma 2-lobed. Mature capsule light brown, broadly ovoid, 0.5-0.6 x 0.4-0.5 cm.

Distribution: Endemic to the Guianas and northern Brazil; on moss-covered dripping rocks (once found on a small tree, acc. Lindeman), on ferrite and granitic boulders in forest, at 100-1100 m alt.; 18 collections studied (SU: 5; FG: 12).

Selected specimens: Suriname: Nassau Mts., Lanjouw & Lindeman 2466 (U); Tafelberg, Hawkins 1888 (MO, US). French Guiana: Région des Emérillons, Mt. Bakra, Cremers 13109 (CAY, B, BR, HAMAB, MO, NY, U, US); Mt. St. Marcel, Sastre 4478 (CAY, P, US).

Phenology: Collected in flower January-April, July, and August, in fruit in April and July.

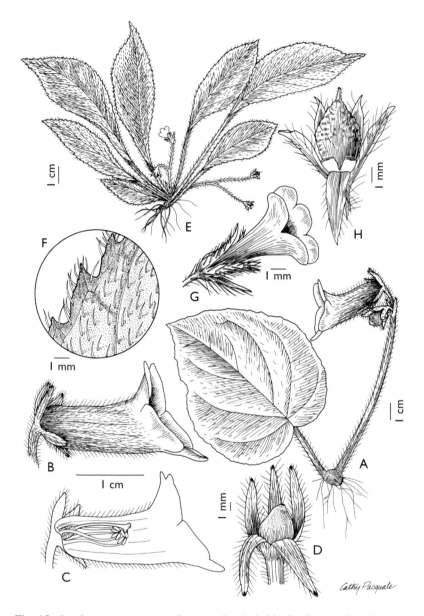

Fig. 15. *Lembocarpus amoenus* Leeuwenb.: A, habit; B, flower; C, calyx and corolla with parts removed to show stamens and pistil; D, young fruit in calyx. *Tylopsacas cuneata* (Gleason) Leeuwenb.: E, habit; F, detail of leaf margin from below; G, flower; H, capsule with persistent calyx and nectary glands. (A, D, Maguire 40788; B-C, Maguire 40806; E-F, Cowan & Soderstrom 1925; G, Hahn *et al.* 4659; H, Maguire *et al.* 32280).

15. **NAPEANTHUS** Gardner, London J. Bot. 2: 13. 1843.
Type: N. brasiliensis Gardner [Napeanthus primulifolius (Raddi) Sandwith]

Terrestrial, caulescent, decumbent herbs, without modified stems. Stems unbranched. Leaves opposite, equal or subequal in a pair, venation pinnate, stomata usually in groups, foliar nectaries absent. Flowers axillary, 1 to many in pedunculate, subumbellate to racemose-paniculate cymes; bracteoles small, persistent, leaf-like; pedicellate. Calyx lobes connate at base up to half their length; corolla white to blue, pink, or lilac, rotate or campanulate, sometimes ventricose; stamens exserted or included, filaments not connate, anthers free, dehiscing by longitudinal slits, thecae divergent; staminode present (when fertile stamens 4); disc absent; ovary superior, stigma slightly 2-lobed. Fruit a dry, loculicidally and septicidally dehiscent, light brown capsule, 4-valved, valves opening slightly. Chromosome number unknown.

Distribution: About 16 species in C and S America from Guatemala southwards to Bolivia, east to the Guianas, and also in southeastern Brazil; in very wet places, often creek banks in rainforests; 4 species in the Guianas.

KEY TO THE SPECIES

1 Leaves linear to narrow-oblanceolate, > 6 times as long as wide, apex acute to acuminate, below purplish . . . . *1. N. angustifolius*
Leaves oblong-spathulate or oblanceolate, < 4 times as long as wide, apex acute to obtuse or rounded, below pale green 2

2 Corolla lobes rounded at apex . . . . *2. N. jelskii*
Corolla lobes emarginate 3

3 Calyx lobes connate for 2 mm at base . . . . *3. N. macrostoma*
Calyx lobes free to base . . . . *4. N. rupicola*

1. **Napeanthus angustifolius** Feuillet & L.E. Skog, Brittonia 54: 358. 2003 ('2002'). Type: French Guiana, Mt. de l'Observatoire, de Granville 6718 (holotype US, isotype CAY, U). – Fig. 17 F-G

Terrestrial herb, 5-25 cm long. Stem subwoody at base, pendent, glabrous or with a few long hairs. Leaves variable; petiole very short or lacking, glabrescent to long pilose; blade thinly membranaceous when dry, linear-oblanceolate, 4-25 x 0.5-2 cm, margin serrate to obscurely serrate, apex acute, base long decurrent to parallel, above with sparse long hairs, below with sparse short hairs. Flowers 2 to many in pedunculate cymes; peduncle 5-15 cm long, with scattered long hairs to glabrescent; pedicel 1-3 cm long, glabrescent. Calyx green, lobes free nearly to base, oblique, equal, lanceolate, 0.5-0.6 x 0.1 cm, margin entire, apex acute with hardened tip, outside with short hairs, inside glabrous; corolla erect in calyx, white, tube campanulate, 0.4-0.5 cm long, base not spurred, ca. 0.1 cm wide, middle not curved or contracted, throat not contracted, lobes subequal, spreading to recurved, wide-lanceolate, margin entire; stamens included, inserted at base of corolla tube; ovary globose, 0.2 x 0.12 cm, glabrous, style 0.3-0.35 cm long, glabrous, stigma navicular, barely swollen. Mature capsule pale brown, ovoid apiculate, 0.2 x 0.12 cm.

Distribution: Endemic to NE French Guiana; on vertical cliffs, deeply shaded, at 0-400 m alt.; 8 collections studied (FG: 8).

Selected specimens: French Guiana: Mts. de Kaw, Billiet & Jadin 6378 (BR, K, MO); Cavernes du Ouanary, Geay 936 (P); Mt. de l'Observatoire, de Granville 6718 (CAY, US).

Phenology: Collected in flower in April and November.

2. **Napeanthus jelskii** Fritsch, Sitzungsber. Akad. Wiss. Wien, Math.-Naturwiss. Kl., Abt. 1. 134: 124. 1925. Type: French Guiana, Cayenne, Jelski s.n. (holotype B destroyed; French Guiana, near Approuague R., Mts. Tortue, Feuillet et al. 10107 (neotype US, here designated, duplicates AAU, B, BBS, BRG, CAY, COL, F, K, INPA, MO, NY, P, PORT, U, VEN, WAG).

Terrestrial herb, 5-10 cm tall. Stem sappy, decumbent or pendent, glabrescent. Leaves variable; petiole very short or lacking, long pilose; blade membranaceous when dry, oblong-spathulate, 1.5-14 x 0.8-4.5 cm, margin remotely serrate, sinuate, or subentire, apex obtuse or rounded, base gradually narrowed towards base, above and below sparsely and minutely pilose. Flowers 2 to many in pedunculate, cymose inflorescences; peduncle 2.5-6 cm long, sparsely pilose; pedicel 1-2.5 cm long, sparsely pilose. Calyx green, lobes connate at base to 1/3 of their length, tube 0.08-0.13 cm, free portion of lobes erect, equal, lanceolate, 0.25-0.4 cm long in bloom, to 0.6 cm in fruit, 0.1 cm wide, margin entire, apex acute, outside pilose, inside glabrous; corolla erect in calyx, white, tube campanulate, 0.3-0.4 cm long, base not spurred, ca. 0.1 cm wide, middle not curved or contracted, throat not contracted, outside pilose, inside glabrous, limb 0.8-1 cm wide, lobes subequal, spreading to recurved, rounded at apex, 0.2 cm long and wide, margin entire; stamens included, inserted at base of corolla tube; ovary globose, 0.1 x 0.1 cm, glabrous, style 0.2-0.25 cm long, glabrous, stigma obscurely saucer-shaped. Mature capsule pale brown, globose, 0.15 x 0.15 cm.

Distribution: Endemic to French Guiana and Amapá, Brazil; on wet rocky boulders, or on rocks along or in the bed of creeks in rainforest, at 50-700 m alt.; 31 collections studied (FG: 30).

Selected specimens: French Guiana: Arataye R., Saut Pararé, Sastre 4823 (CAY, P), 5699 (CAY, P, US); Mts. Tortue, Cr. Tawen, Feuillet 9882 (CAY, E, US); Mt. Bellevue de l'Inini, de Granville et al. 7881 (B, CAY, P, U, US); Mts. Françaises, Sastre & Bell 8055 (CAY, P, US); near Approuague R., Mts. Tortue, Feuillet et al. 10107 (AAU, B, BBS, BRG, CAY, COL, F, K, INPA, MO, NY, P, PORT, U, US, VEN, WAG).

Phenology: Collected in flower throughout the year, in fruit in September and November.

Note: For selecting the neotype two main criteria were used: - many duplicates are deposited on three continents, - three specialists of the Gesneriaceae, Leeuwenberg and both authors, agreed on the identity on the specimens.

3. **Napeanthus macrostoma** Leeuwenb., Acta Bot. Neerl. 13: 64. 1964. Type: Brazil, Amapá, Egler & Irwin 46401 (holotype NY, isotypes U, US, WAG). – Fig. 16

Terrestrial herb, 10-25 cm tall. Stem subwoody at base, decumbent or pendent, pilose, becoming glabrous. Leaves variable; petiole very short or lacking, pilose; blade membranaceous when dry, oblong-spathulate, 11-17 x 3.2-5.6 cm, margin sinuate or shallowly crenate-serrate, apex obtuse or rounded, base gradually narrowed towards base, above sericeous when unfolding, with thin cobweb-like hairs soon glabrescent, below puberulous with thin appressed hairs on midrib and veins. Flowers several, in pedunculate, subumbellate inflorescences; peduncle 2.5-6 cm long, with spreading thin hairs; pedicel 0.5-1 cm long, with spreading thin hairs. Calyx green, lobes erect, equal, lanceolate, 0.8 x 0.2 cm, connate for about ¼ of their length, margin entire, apex acute, outside with thin hairs, inside glabrous; corolla erect in calyx, white, tube campanulate, ca. 0.8 cm long, base not spurred, 0.15-0.2 cm wide,

72

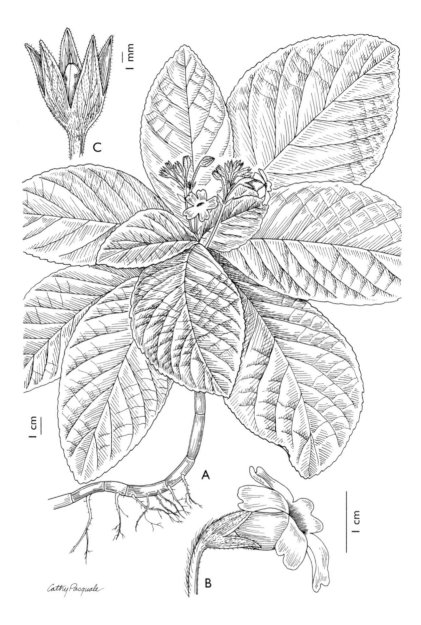

Fig. 16. *Napeanthus macrostoma* Leeuwenb.: A, habit; B, flower; C, fruit.
(A-B, from photo of de Granville *et al.* 8482; C, de Granville *et al.* 7719).

middle not curved or contracted, throat not contracted, outside glabrous, inside minutely pubescent, limb 1-1.2 cm wide, lobes subequal, spreading to recurved, ovate-rectangular, 0.4 cm long, 0.3 cm wide, margin minutely ciliate, emarginate; stamens included, inserted at base of corolla tube; ovary ovoid, 0.15 x 0.1 cm, glabrous, style 0.3 cm long, glabrous, stigma obscurely saucer-shaped. Mature capsule pale brown, ovoid, 0.3 cm long, 0.2 cm wide.

Distribution: Suriname, French Guiana, and Brazil (Amapá); on vertical rocks, or on steep slopes or river banks in rainforest, at 70-750 m alt.; 55 collections studied (SU: 5; FG: 50).

Selected specimens: Suriname: Nassau Mts., Lanjouw & Lindeman 2630 (U, US); near Tafelberg, Wessels Boer 1577 (U, US). French Guiana: Mt. Atachi Bacca, de Granville *et al.* 10729 (CAY, B, NY, P, US); Mt. Bellevue de l'Inini, de Granville *et al.* 7719 (B, BR, CAY, INPA, MG, MO, P, U, US).

Phenology: Collected in flower throughout the year, in fruit in February, March, August, November, & December.

Note: Photograph: Feuillet & Skog, 2002 (pl. 65 e (Mori *et al.* 21657)).

4. **Napeanthus rupicola** Feuillet & L.E. Skog, Brittonia 54: 360. 2003 ('2002'). Type: Guyana, Potaro-Siparuni Region, Mt. Wokomung, Boom & Samuels 9202 (holotype US, isotype NY). – Fig. 17 A-E

Terrestrial herb, 10-30 cm tall. Stem subwoody at base, decumbent or pendent, strigose. Leaves variable; petiole very short or lacking long, appressed pubescent; blade chartaceous when dry, oblanceolate, 7-15 x 3-6 cm, margin obscurely serrate, apex acute to obtuse, base long decurrent, above appressed pubescent along veins, below shortly to densely pubescent. Flowers few to many in pedunculate cymes; peduncle 1.5-4 cm long, pilose; pedicel 2-5 cm long, short pubescent. Calyx green, lobes free, erect, equal, lanceolate, 0.7-1 x 0.25 cm, margin entire, apex acute with hardened tip, outside hirsute, inside puberulous; corolla erect in calyx, white, tube campanulate, 0.5-0.7 cm long, base not spurred, 0.2 cm wide, middle not curved or contracted, throat slightly constricted, outside lower tube glabrous, inside glabrous, limb 1.1 cm wide, lobes subequal, spreading to recurved, obovate, 0.3-0.4 x 0.3 cm, margin minutely ciliate, emarginate; stamens included, inserted at base of corolla tube; ovary globose, 0.4-0.5 x 0.3-0.4 cm, glabrous, style 0.2-0.3 cm long, glabrous, stigma slightly capitate. Mature capsule pale brown, ovoid, 0.5-0.6 x 0.4 cm.

74

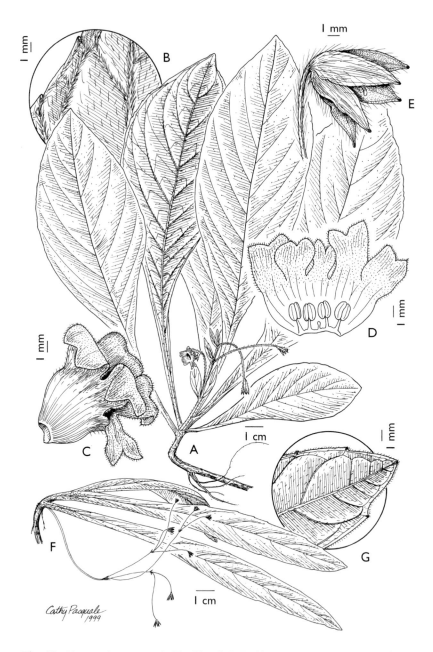

Fig. 17. *Napeanthus rupicola* Feuillet & L.E. Skog: A, habit; B, detail of leaf
margin; C, corolla. D, corolla opened, with stamens and staminode; E, calyx.
Napeanthus angustifolius Feuillet & L.E. Skog: F, habit; G, detail of leaf margin.
(A-E, Hahn *et al.* 4326; F-G, Herb. Maire s.n.).

Distribution: Endemic to Guyana, Potaro-Siparuni and Cuyuni-Mazaruni Regions; moist habitats, on rocks in montane and cloud forests, at 1100-1600 m alt.; 3 collections studied (GU: 3).

Specimens studied: Guyana: Potaro-Siparuni Region, Mt. Kopinang, Hahn *et al.* 4326 (BRG, US); Mt. Wokomung, Boom & Samuels 9202 (NY, US); Cuyuni-Mazaruni Region, Pakaraima Mts., slopes on NW side of Mt. Ayanganna, Henkel & Hoffman 169 (BRG, US).

Phenology: Collected in flower in February and November.

16. **NAUTILOCALYX** Linden ex Hanst., Linnaea 26: 207. 1854 ('1853'), nom. cons. – *Episcia* Mart. sect. *Nautilocalyx* (Linden ex Hanst.) Benth. & Hook. f., Gen. Pl. 2: 1007. 1876.
Type: N. hastatus Linden ex Hanst., nom. illeg. (Centrosolenia bractescens Hook.) [Nautilocalyx bracteatus (Planch.) Sprague]

Centrosolenia Benth., Lond. J. Bot. 5: 362. 1864, nom. rej. – *Episcia* sect. *Centrosolenia* (Benth.) Benth. & Hook. f., Gen. Pl. 2: 1007. 1876. – *Episcia* subsect. *Centrosolenia* (Benth.) Leeuwenb., Acta Bot. Neerl. 7: 310. 1958.
Type: C. hirsuta Benth. [Nautilocalyx cordatus (Sprague) L.E. Skog].
Episcia sect. *Trichosperma* Leeuwenb., Acta Bot. Neerl. 7: 312. 1958.
Type: E. bryogeton Leeuwenb.

Terrestrial or rarely epiphytic, caulescent, decumbent to erect herbs or low shrubs, occasionally tuberous. Stems rarely branched. Leaves opposite, equal or unequal in a pair, venation pinnate, foliar nectaries absent. Flowers axillary, 1-15 in pedunculate or epedunculate, fasciculate, or cymose inflorescences; bracteoles usually present; pedicellate. Calyx lobes free; corolla white to yellow, red or purple, usually with spots or lines of purple, tubular, broadened laterally; stamens included, filaments basally connate, anthers apically coherent in 2 pairs, dehiscing by longitudinal slits, thecae parallel or divergent; staminode minute to small; disc a single dorsal gland or 2 opposite glands; ovary superior, stigma stomatomorphic to 2-lobed. Fruit a fleshy, loculicidally dehiscent, white or colored capsule, 2-valved, valves opening slightly or to 180°.
Chromosome number n=9 (18) (Skog 1984).

Distribution: More than 50 species found in the Lesser Antilles, C America from Mexico to Panama, and throughout northern S America from Peru to Colombia, to Venezuela and the Guianas, and Amazonian Brazil; in or at the edge of rainforest, occasionally in seasonally dry locations; 11 species in the Guianas.

Notes: The generic name derives from a supposed similarity between the shape of the bracts of the type species and a nautilus shell (fide Planchon).

The presence of *N. villosus* (Kunth & Bouché) Sprague had been suggested in French Guiana (Leeuwenberg, 1958). This species seems to be restricted to western Venezuela and neighbouring parts of Colombia. All specimens from the Guianas identified as *N. villosus* can be placed in *N. fasciculatus*, *N. pallidus*, or *N. pictus*.

A few specimens document the presence of *N. melittifolius* (L.) Wiehler (mentioned as *Episcia melittifolia* (L.) Mart. in Leeuwenberg 1958, 1984) as a house and garden ornamental plant in French Guiana. It is not known to escape cultivation and we include it only in the key.

KEY TO THE SPECIES

1   Corollas red 2
     Corollas white, or slightly tinged with pink (tube white with lobes lavender
        blue in *N. cordatus*) 6

2   Flower 1 per axil . . . . *6. N. kohlerioides*
     Flowers 2 to several per axil (rarely 1 in *N. porphyrotrichus*) 3

3   Calyx lobes rounded at apex . . . . *3. N. coccineus*
     Calyx lobes acute 4

4   Inflorescences pedunculate; calyx lobes sparsely villous or glabrous outside
        (cultivated) . . . . *N. melittifolius*
     Inflorescences epedunculate; calyx lobes hirsute or densely villous outside . 5

5   Leaf blades subequal in a pair; corolla oblique in calyx . . *5. N. fasciculatus*
     Leaf blades unequal in a pair; corolla erect in calyx . . . .
        . . . . *10. N. porphyrotrichus*

6   Leaf blades decurrent into petiole 7
     Leaf blades not decurrent 9

7   Leaf blades purplish underneath . . . . *11. N. punctatus*
     Leaf blades green underneath 8

8   Calyx lobes entire . . . . *7. N. mimuloides*
     Calyx lobes repand-serrate . . . . *8. N. pallidus*

9   Leaf blades acuminate at apex, cuneate at base . . . . *9. N. pictus*
     Leaf blades obtuse or acute at apex, rounded or subcordate at base 10

10 Calyx lobes with a few teeth; corolla tube 0.8-1.1 cm long . . . .
    . . . . 2. *N. bryogeton*
    Calyx lobes serrate; corolla tube 2.8-5 cm long 11

11 Leaf blades equal in a pair, ovate, below minutely pilose; corolla tube white,
    2.8 cm long, lobes white . . . . *1. N. adenosiphon*
    Leaf blades mostly unequal in a pair, oblong-elliptic, below hirsute; corolla
    tube white, 3.3-5 cm long, lobes lavender . . . . *4. N. cordatus*

1. **Nautilocalyx adenosiphon** (Leeuwenb.) Wiehler, Selbyana 5: 29.
    1978. – *Episcia adenosiphon* Leeuwenb., Acta Bot. Neerl. 18: 858.
    1969. Type: Venezuela, Bolívar, Steyermark 88162 (holotype WAG,
    isotypes VEN, WAG).

Terrestrial herb, 10-20 cm tall. Stem sappy, creeping or ascending,
tomentose at apex, glabrescent below. Leaves equal in a pair; petiole
1-3 cm long, pilose or villous; blade chartaceous when dry, ovate or
nearly so, 2-4 x 1.2-2.8 cm, margin crenate-serrate, apex obtuse or acute,
base rounded (sometimes subcordate in cultivated specimens) above
appressed-pubescent, below minutely pilose. Flowers solitary;
epedunculate; pedicel 1.5 cm long, villous. Calyx pale green, lobes
connate at very base, 4 erect, dorsal one curved around spur, subequal,
ovate or narrowly ovate, 1.1-1.2 x 0.4-0.6 cm, margin serrate, apex
acuminate, outside pilose, inside glabrous; corolla slightly oblique in
calyx, white, 3.5 cm long, tube 2.8 cm long, base spurred, ca. 0.5 cm
wide, middle nearly cylindrical, throat not contracted, ca. 1 cm wide,
outside pilose, inside pubescent all over, limb ca. 2 cm wide, lobes
subequal, spreading, suborbicular, 0.7-0.8 x 0.7-0.8 cm, margin entire;
stamens included, inserted near base of corolla tube; ovary ovoid, 0.3 x
0.2 cm, densely pubescent, style 1.2 cm long, sparsely pilose-pubescent,
stigma 2-lobed. Mature fruit not seen.

Distribution: Venezuela (Estado Bolívar, Sierra Imataca), French
Guiana; in river side forest, at 60-420 m alt.; 7 collections studied (FG: 5).

Specimens studied: French Guiana: Basse Mana R., Saut Fracas,
de Granville (4624); bassin du Sinnamary, Camp Eugène, Cremers *et
al.* 13766 (CAY, P, US); Haute Mana R., village Bellevue, Cremers
7567 (CAY); Mt. Lucifer, de Granville *et al.* 13959 (CAY, US), Blanc
85-124 (CAY).

Phenology: Collected in flower in February and June-December, in
fruit in November.

2. **Nautilocalyx bryogeton** (Leeuwenb.) Wiehler, Selbyana 5: 30. 1978.
  – *Episcia bryogeton* Leeuwenb., Acta Bot. Neerl. 7: 312, 400, fig. 22.
  1958. Type: Guyana, Pinkus 12 (holotype NY, isotype GH, US).

Terrestrial suffrutescent herb, 2-50 cm long. Stem subwoody towards base, creeping, strigillose at apex, glabrescent below. Leaves strongly unequal in a pair; petiole (0.4-)1-2.5(-4) cm long, villous; blade papyraceous when dry, oblong-ovate or oblanceolate, narrowed towards base, larger blade 3.2-9.5 x 1.7-4.5 cm, margin crenate-serrate, apex acute or obtuse, base rounded or subcordate, above tomentose, below strigillose, especially on midrib and veins. Flowers 3-10, in fascicles; pedunculate; pedicel 0.6-1.7 cm long, villous. Calyx green, lobes free, 4 erect, subequal oblong-lanceolate, narrowed towards base, 0.4-0.6 x 0.1-0.15 cm, dorsal one curved around spur, somewhat smaller and narrower, margin with some teeth, apex acute, outside villous, inside villous above and glabrous below; corolla oblique in calyx, white, 1-1.4 cm long, tube infundibuliform, 0.8-1.1 cm long, base shortly spurred, 0.15 cm wide, middle gradually widened to throat, slightly bent downwards, throat not ventricose, 0.4 cm wide, outside villous, inside glabrous, limb 1-1.1 cm wide, lobes subequal, spreading, rounded, 0.3-0.4 x 0.3-0.4 cm, margin entire; stamens included, inserted on base of corolla; ovary globose, 0.1 x 0.1 cm, softly hirsute, style ca. 0.6 cm long, glabrous, stigma stomatomorphic. Mature capsule globose, 0.3 x 0.3 cm.

Distribution: Endemic to Guyana; on moss-covered rocks, on slopes, at 1000 m alt.; 5 collections studied (GU: 5).

Specimens studied: Guyana: Cuyuni-Mazaruni Region, Pakaraima Mts., Kurupung R., near Makreba Falls, Hoffman 2073 (US), Pinkus 12 (GH, NY, US); Meamu R. headwaters, Hoffman 2263 (US); Membaru-Kurupung trail, Alston 371 (K(2)), Maguire & Fanshawe 32392 (NY, US).

Phenology: Collected in flower May and November.

3. **Nautilocalyx coccineus** Feuillet & L.E. Skog, Brittonia 54: 352. 2003
  ('2002'). Type: Guyana, Upper Potaro R. Region, upper slopes of Mt. Wokomung, Boom & Samuels 9186 (holotype US, isotype NY).
                                                                      – Fig. 18

Terrestrial herb, 5-10 cm long. Stem sappy, prostrate, hirsute. Leaves unequal in a pair; petiole 1.5-10 cm long, hirsute; blade chartaceous when dry, obovate, larger blade 6-15 x 3-6 cm, margin crenate to widely

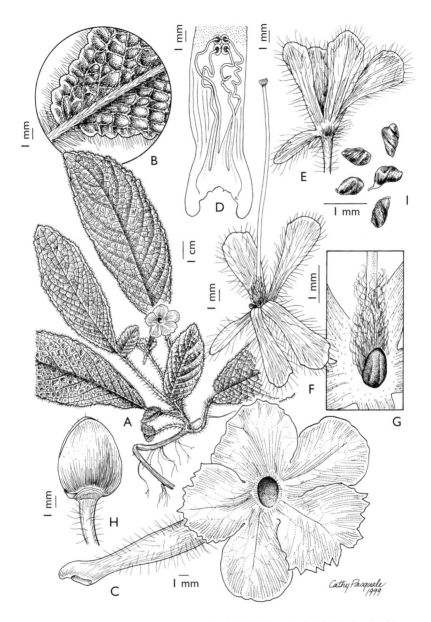

Fig. 18. *Nautilocalyx coccineus* Feuillet & L.E. Skog: A, habit; B, detail of lower leaf blade base; C, corolla; D, corolla opened to show stamens and rudimentary staminode; E, calyx; F, nectary gland and pistil in opened calyx; G, detail of nectarial gland and ovary; H, capsule (calyx removed); I, seeds. (A, E-F, Pipoly *et al*. 11085; B, H-I, Tillett *et al*. 44947; C-D, G, Boom *et al*. 9186).

serrate, apex round to obtuse, blunt, base asymmetrically acute to rounded, above hirsute-floccose, below hirsute on veins. Flowers 2-8, in epedunculate or short-pedunculate cymes; peduncle 0.2-0.5 cm long, villous; pedicel 1-3.5 cm long, hirsute. Calyx green, lobes free, 4 erect, subequal, spathulate, 1 x 0.3-0.4 cm, dorsal one spreading, smaller, margin with 2 lateral teeth near apex, apex rounded, outside and inside hirsute; corolla strongly oblique in calyx, red, 3-4 cm long, tube straight narrow tubular, 1.7-3 cm long, base spurred, 0.2 cm wide, middle cylindric, throat widening, 0.2-0.3 cm wide, outside hirsute, inside glabrous in its basal part, puberulent in upper half, limb 2-2.5 cm wide, lobes subequal, spreading, suborbicular, 0.5-0.6 x 0.5-0.7 cm, margin entire to serrate; stamens included, inserted near base of corolla; ovary conical, 0.15-0.2 x 0.1 cm, with soft wavy pubescence, style 1.5-2 cm long, glabrous, stigma obscurely 2-lobed. Mature capsule light brown, globose (probably somewhat flattened), 0.6 x 0.5 cm.

Distribution: Endemic to Guyana in the Upper Potaro and Cuyuni-Mazaruni Regions, near Mt. Ayanganna and Wokomung; on rocks in moist forest, at 700-1600 m alt.; collections studied (GU: 5).

Specimens studied: Guyana: Cuyuni-Mazaruni Region: Mt. Ayanganna, Tillett *et al.* 44947 (NY, US), Maguire *et al.* 40586a (US), Clarke *et al.* 9256 (US; headwaters of W branch of Kangu R., near Mt. Ayanganna, Pipoly *et al.* 11085 (BRG, US); Upper Potaro R. Region, upper slopes of Mt. Wokomung, Boom & Samuels 9186 (NY, US).

Phenology: Collected in flower in March and July, in fruit in March.

4. **Nautilocalyx cordatus** (Gleason) L.E. Skog in L.E. Skog & Steyerm., Novon 1: 217. 1991. – *Episcia cordata* Gleason, Bull. Torrey Bot. Club 58: 466. 1931. Type: Venezuela, Amazonas, Tate 878 (holotype NY, isotypes K, US). – Fig. 19 A-F

Centrosolenia hirsuta Benth., London J. Bot. 5: 362. 1846. – *Episcia hirsuta* (Benth.) Hanst., Linnaea 34: 350. 1865 ('1865-1866') (non *Nautilocalyx hirsutus* (Sprague) Sprague 1912). Type: Guyana, Banks of R. Parama, Ro. Schomburgk s.n. (holotype K).

Terrestrial herb, <10 cm tall. Stem sappy, creeping, hirsute at apex, glabrescent below. Leaves subequal to strongly unequal in a pair; petiole 0.7-13 cm long, more or less hirsute; blade chartaceous when dry, often obliquely oblong-elliptic, variable in shape and size, blade of the larger leaf in a pair 2-15 x 1.3-8 cm, margin irregularly crenate-serrate (larger leaves often bicrenate), apex acute, obtuse or rounded, base subcordate

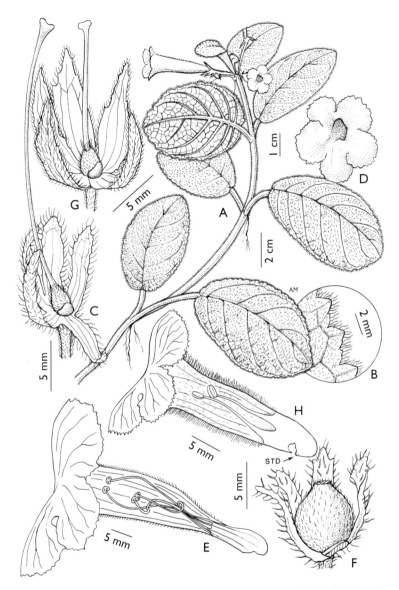

Fig. 19. *Nautilocalyx cordatus* (Gleason) L.E. Skog: A, habit; B, leaf margin; C, calyx showing nectary and pistil; D, corolla, front view; E, corolla with parts removed to show stamens; F, calyx with parts removed to show young fruit. *Nautilocalyx kohlerioides* (Leeuwenb.) Wiehler: G, calyx showing nectary and pistil; H, corolla with parts removed to show stamens and staminode [STD]. (A-B, Pipoly *et al.* 10449 + Steyermark *et al.* 124076 [Venezuela]; C-E, Gillespie 960; F, Pipoly *et al.* 10537; G-H, Jacquemin 1616).

82

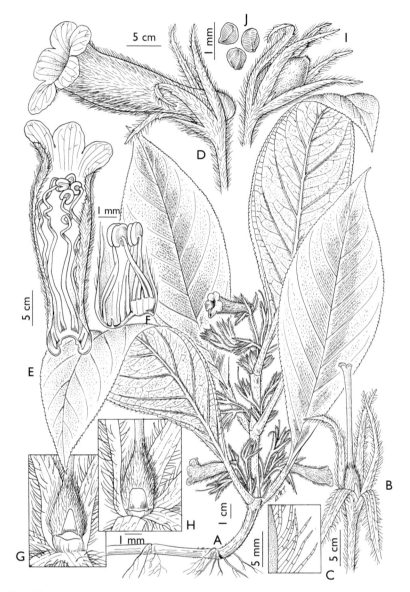

Fig. 20. *Nautilocalyx fasciculatus* L.E. Skog & Steyerm.: A, habit; B, flower with corolla removed; C, calyx lobe margin; D, flower; E, corolla opened showing old stamens; F, corolla base with young stamens; G, ventral nectariferous gland; H, dorsal nectariferous gland; I, capsule with persistent calyx; J, seeds. (A-J, Maguire *et al.* 29982). Reproduced with permission from L.E. Skog & J. Steyermark, Novae Gesneriaceae Neotropicarum III. Additional new species from Venezuela. Novon 1: 211-222. 1991.

or rounded, above hirsute, below hirsute, especially on veins. Flowers 2-6, in epedunculate fascicles; pedicel 1-5 cm long, hirsute. Calyx green or purplish, lobes free, 4 erect, subequal, lanceolate, 0.6-1.5 x 0.12-0.5 cm, margin serrate, apex acuminate, outside and inside hirsute, dorsal one curved around spur, somewhat shorter and narrower; corolla oblique in calyx, white outside, inside whitish with lavender lobes, 4.7-7 cm long, tube trumpet-shaped, 3.3-5 cm long, base spurred, 0.25-0.5 cm wide, middle dorsally slightly ventricose, slightly bent downwards, throat contracted, 1-1.5 cm wide, outside villous, inside glabrous, limb 1.5-4.2 cm wide, lobes subequal, spreading, rounded, 0.6-1.7 x 0.6-1.7 cm, margin crenate-serrate; stamens included, inserted on base of corolla; ovary ovoid, 0.3-0.4 x 0.15-0.2 cm, softly hirsute, style ca. 3 cm long, glabrous or nearly so, stigma obscurely 2-lobed. Mature capsule light brown, globose, 0.7-0.8 x 0.7-0.8 cm.

Distribution: Venezuela, Guyana, Suriname, and northern Brazil (Amapá, Amazonas, Pará, Roraima); on wet rocks near waterfalls in montane rainforest, at 170-1500 m alt.; 55 collections studied (GU: 40; SU: 5).

Selected specimens: Guyana: Banks of Parama R., Ro. Schomburgk s.n. (K); Kaieteur National Park, Pipoly et al. 9951 (CAY, FDG, NY, U, US); Mt. Ayanganna, Maguire et al. 40586b (NY, US). Suriname: Tumuc-Humac, Cr. Grand Koulé-Koulé, Sastre 1469 (CAY, P), 1470 (CAY, P).

Phenology: Collected in flower throughout the year, 1 collection recorded in fruit, in May.

5. **Nautilocalyx fasciculatus** L.E. Skog & Steyerm., Novon 1: 219. 1991. Type: Venezuela, Amazonas, Maguire, Cowan & Wurdack 29982 (holotype US, isotype US). – Fig. 20

Terrestrial herb, 20-50 cm tall. Stem sappy, erect, decumbent at base, sparsely villose. Leaves subequal in a pair; petiole 2-3 cm long, sparsely villous; blade membranous when dry, oblong-elliptic, 11-18 x (3.5-)4.5-7 cm, margin closely crenate-serrulate, apex acute, base acute, above and below sparsely villous. Flowers 2-6, in epedunculate fascicles; pedicel 0.8-2 cm long, densely villous. Calyx lobes connate at base, erect, unequal, lanceolate to linear-lanceolate, 1-1.5 x 0.2 cm, margin with 0-4 thickened teeth, apex narrowly acute, outside densely villous, inside sparsely villous; corolla oblique in calyx, red, 3.5-5.5 cm long, tube narrowly infundibuliform, 3.3-5 cm long, base spurred, 0.3-0.4 cm wide, middle slightly enlarged, throat not contracted, 0.5 cm wide, outside densely villous except at base, inside glabrous, limb 1.5 cm wide, lobes subequal, spreading, suborbicular, 0.5-0.6 x 0.5-0.6 cm, margin wavy-serrate; stamens included, inserted at base of corolla, adnate for 0.9 cm; ovary narrowly ovoid, 0.45 x 0.2 cm, densely sericeous, style 1.8-2 cm long, glabrous, stigma 2-lobed. Mature capsule ovoid, 2 x 1 cm.

Distribution: Venezuela (Amazonas) and the Guianas; in damp forest, 200-400 m alt.; 12 collections studied (GU: 5; SU: 3; FG: 2).

Selected specimens: Guyana: Acarai Mts., Henkel *et al.* 4829 (US); Wassarrai Mts., Clarke 8564 (US). Suriname: Haut Litany, de Granville *et al.* 12100 (CAY, US); Tumuc-Humac, E of Paloulouiméenpeu, de Granville 1104 (CAY). French Guiana: Tumuc-Humac, between Mitaraka and Suriname border, de Granville 1118 bis (CAY); Mts. Mitaraka, de Granville B-4530 (CAY, US).

Phenology: Collected in flower in August.

6. **Nautilocalyx kohlerioides** (Leeuwenb.) Wiehler, Selbyana 5: 34. 1978. – *Episcia kohlerioides* Leeuwenb., Acta Bot. Neerl. 13: 61, fig. 2. 1964. Type: Brazil, Amapá, Oiapoque R., Irwin *et al.* 48315 (holotype NY, isotype U, US).   – Fig. 19 G-H

Terrestrial herb, about 15-30 cm tall. Stem sappy, creeping, ochraceous-tomentose. Leaves equal or unequal in a pair; petiole 1-4.5 cm long, ochraceous-tomentose; blade chartaceous when dry, ovate, oblong-ovate or oblong-elliptic, blade of the larger leaf in a pair 7-12 cm long, 4-6 cm wide, margin crenate-serrate, apex acuminate, base rounded and often unequal-sided, above strigillose, below villous. Flowers solitary; epedunculate; pedicel 1.5-4 cm long, villous. Calyx green, with pinkish pubescence when in fruit, lobes connate at base, tube 0.1 cm, free portion of lobes 4 erect, subequal, leafy, lanceolate, 1-2.2 x 0.3-0.5 cm, dorsal one curved around spur, somewhat smaller and narrower, margin serrate, apex acuminate, outside villous, inside pubescent except for villous apex; corolla oblique in calyx, bright red, 2.7-3.2 cm long, tube cylindric, 2.2-2.5 cm long, base shortly spurred, 0.5 cm wide, middle slightly widened, throat slightly contracted, 0.5-0.9 cm wide, outside densely villous, inside pilose in throat, limb 1.5-2 cm wide, lobes subequal, spreading, nearly orbicular, 0.6-1 x 0.6-1 cm, margin entire or obscurely toothed; stamens included, inserted at about 0.2 cm from base of corolla; ovary ovoid, 0.4 x 0.3-0.5 cm, hirsute, style 0.8-1.7 cm long, glabrous, stigma saucer-shaped. Mature capsule globose, 1 x 1 cm.

Distribution: Brazil (Amapá) and French Guiana; on moist places in rain forest, at 140-200 m alt.; 23 collections studied (FG: 20).

Selected specimens: French Guiana: Oyapock R., un peu en amont du Saut Kouamantapéré, Oldeman T-667 (CAY, P, U, US); Tamouri R., Lescure 119 (CAY, P, US); Trois Saut, Oldeman T-948 (CAY, P, US).

Phenology: Collected in flower in March, May, August, and September.

Vernacular names: French Guiana: yamulepila (Wayãpi; Jacquemin 1523); ka'iuwitoto, yamuleka'apilã (Wayãpi; Grenand *et al.*, 1987).

Use: The Wayãpi Amerindians provided different informations: some of them called yamuleka'apilã both *Nautilocalyx kohlerioides* and *Columnea calotricha*, and used the leaves of either one crumpled and macerated in water as febrifuge in external wash, or against headaches in poultice rubbed on the forehead. Others used only *Nautilocalyx kohlerioides* called ka'iuwitoto. The flowers and leaves were used in decoction or maceration, as external wash for the babies at risk of declining health because their father transgressed a hunting taboo concerning *Cebus apella* (Grenand *et al.*, 1987).

7. **Nautilocalyx mimuloides** (Benth.) C.V. Morton, Fl. Trinidad Tobago 2: 304. 1954. – *Episcia mimuloides* Benth., London J. Bot. 5: 362. 1846. Type: Venezuela, Roraima, Ro. Schomburgk, ser. II 843 (lectotype K, isolectotypes BR, GM, G, P, P, W) (designated by Leeuwenberg 1958: 409).

Terrestrial herb, 15-40 cm tall. Stem sappy, creeping or ascending, puberulous at apex, glabrescent. Leaves equal or subequal in a pair; petiole 1-4.5 cm long, puberulous; blade membranous when dry, oblanceolate to elliptic, 5-19 x 2.5-7.5 cm, margin crenate-serrate, apex acuminate, base decurrent into petiole, above pubescent, below puberulous, especially on midrib and veins. Flowers 2-8, umbellate or subcymose; peduncle 0.5-1.5 cm long, puberulous; pedicel 1-3 cm long, puberulous to glabrous. Calyx green, lobes free, 4 erect, dorsal lobe curved around spur, subequal, ovate or oblong-ovate, 1-1.6 x 0.5-0.7 cm, margin entire, apex acuminate, outside sparsely pubescent, inside minutely puberulous; corolla oblique in calyx, white (rose-white, acc. Archer 2321), 4-6 cm long, tube infundibuliform, 3-4.5 cm long, base spurred, 0.4-0.5 cm wide, middle slightly widened, hardly ventricose, throat barely contracted, 0.9-1.4 cm wide, outside pilose, inside with scattered minute 2-celled glandular hairs, limb 1.7-2.6 cm wide, lobes subequal, spreading, rounded, 0.4-0.6 x 0.4-0.6 cm, margin entire; stamens included, inserted on corolla 0.3-0.4 cm from base; ovary ovoid, 0.3-0.4 x 0.2-0.3 cm wide, hirsute, style 2.5-3 cm long, glabrous, stigma 2-lobed. Mature capsule globose, 1 x 1 cm.

Distribution: Tobago, the Guianas, and Brazil (Amapá); in rainforests and disturbed bare edges, at 10-840 m alt.; 100 collections studied (GU: 20; SU: 1; FG: 45).

Selected specimens: Guyana: Barama R., van Andel *et al.* 1075 (U, US); Pakaraima Mts., Paruima Mission, Maas *et al.* 5599 (BBS, U, US). Suriname: Inselberg Talouakem, Massif des Tumuc-humac, de Granville *et al.* 12107 (BBS, CAY, US). French Guiana: Sommet Tabulaire, 50 km SE de Saül, de Granville 3688 (CAY, MO, P, U, US); Mts. de La Trinité, Inselberg Nord Ouest, de Granville *et al.* 6148 (CAY, U, US).

Phenology: Collected in flower January-October and in fruit in May.

Note: Photographs: Feuillet & Skog, 2002 (pl. 66 a (Mori *et al.* 22212), pl. 66 b (Mori *et al.* 23049)).

8. **Nautilocalyx pallidus** (Sprague) Sprague, Bull. Misc. Inform. Kew 1912: 89. 1912. – *Alloplectus pallidus* Sprague, Bull. Misc. Inform. Kew 1911: 346. 1911. Type: Cult. Hort. Kew. (holotype K), originally from Peru, Forget s.n.

Terrestrial herb, 10-30 cm tall. Stem sappy, erect, decumbent at base, with long scattered hairs. Leaves subequal in a pair; petiole 1-3 cm long, with scattered appressed hairs; blade membranous when dry, lanceolate to elliptic, 6-30 x 2-8 cm, margin slightly serrate, apex acuminate, base decurrent, above with scattered hairs, mostly on veins, below with long scattered hairs, shorter and denser on veins. Flowers 2-many, in epedunculate fascicles; pedicel 0.5-1 cm long, minutely pubescent. Calyx pale green, lobes connate at base, tube 0.2-0.3 cm, free portion of lobes oblique to erect, subequal, lanceolate bracteiform, 1-2 x 0.5-1 cm, margin repand-serrate, apex acute to slightly acuminate, outside and inside glabrescent; corolla erect in calyx, white, 3-5 cm long, tube cylindric, 2.5-4.5 cm long, base spurred, 0.3-0.4 cm wide, middle cylindric, throat round, 1 cm wide, outside densely hairy, inside glabrescent, limb 3 cm wide, lobes subequal, spreading, rounded, 0.7-1.1 x 1.2-1.3 cm, margin ciliate; stamens included, inserted 0.5 cm from base of corolla; ovary ovoid, 0.5 x 0.2-0.3 cm, pilose, style 3 cm long, hirsute, stigma 2-lobed. Mature fruit not seen.

Distribution: Amazonia (Colombia, Ecuador, Peru, Bolivia, Brazil, Venezuela) and the Guianas; ripicolous forest, at low elevations; > 50 collections studied (FG: 4).

Selected specimens: French Guiana: R. Oyapock, Cr. Armontabo, Grenand & Prévost 1996 (CAY), Prévost 1909 (CAY, P, U, US).

Phenology: Collected in flower in April-May.

9. **Nautilocalyx pictus** (Hook.) Sprague, Bull. Misc. Inform. Kew 1912: 88. 1912. – *Centrosolenia picta* Hook., Bot. Mag. 77: ad pl. 4611. 1851. – *Collandra picta* (Hook.) Lem., Jardin Fleuriste 2: ad pl. 214. 1852. – *Episcia picta* (Hook.) Hanst. in Mart., Fl. Bras. 8(1): 403. 1864. – *Columnea picta* (Hook.) Hanst. in Mart., Fl. Bras. 8(1): 422. 1864. Type: Cult. Hort. Kew. (holotype K), originally from Brazil, collected by Spruce.   – Fig. 21

Nautilocalyx lacteus Sandwith, Bull. Misc. Inform. Kew 1931: 489. 1931. Type: Guyana, Sandwith 3 (holotype K).

Terrestrial herb, 20-50 cm tall. Stem sappy, creeping or ascending, purple-villous or pilose at apex, glabrescent. Leaves subequal in a pair; petiole (1-)1.5-5(-8) cm long, villous or pilose; blade chartaceous when dry, oblong-elliptic or oblong-lanceolate, 5.5-22.5 x 2.5-10 cm, margin crenate-serrate, apex acuminate, base cuneate, above sparsely strigose to glabrous, below appressed-pubescent or pilosulous, especially on midrib and veins. Flowers 1 to several; peduncle obsolete or very short; pedicel 0.3-1.5 cm long, purple villous. Calyx green or purple, lobes almost free, 4 erect, subequal, linear-lanceolate, 1-2.5 x 0.15-0.4 cm, dorsal one curved around spur, somewhat smaller, margin with few thickened, hard teeth, apex long-acuminate, outside and inside densely villous; corolla oblique in calyx, white, 3-4.5 cm long, tube more or less infundibuliform, 2.5-3.5 cm long, base spurred, 0.2-0.4 cm wide, middle slightly contracted, throat slightly contracted, 0.7-1.1 cm wide, outside more or less pilose, inside dorsally pubescent with glandular hairs, limb 1.5-2 cm wide, lobes subequal, 2 dorsal lobes somewhat smaller than others, spreading, rounded, 0.4-0.7 x 0.4-0.7 cm, margin more or less serrate; stamens included, inserted at base of corolla; ovary ovoid, 0.25-0.4 x 0.15-0.3 cm, hirsute, style 2-2.5 cm long, pubescent, stigma 2-lobed. Mature capsule light brown, globose, 0.6-0.8 x 0.6-0.8 cm.

Distribution: Colombia, the Guianas and northern Brazil; in rainforests, at 20-1050 m alt.; > 200 collections studied (GU: 25; SU: 15; FG: 165).

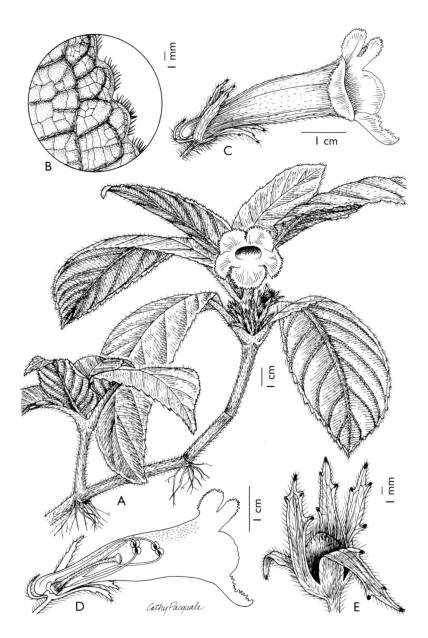

Fig. 21. *Nautilocalyx pictus* (Hook.) Sprague: A, habit; B, leaf margin; C, flower; D, flower with parts of calyx and corolla removed to show stamens, nectary, and pistil; E, young fruit in calyx. (A, from photo; B, de Granville 8657; C-D, Brothers 7605 [cultivated plant, origin French Guiana]; E, Oldeman B-534).

Selected specimens: Guyana: Kangaruma-Potaro Landing, Gleason 199 (NY); North Fork R., McDowell *et al.* 4823 (U, US). Suriname: Tafelberg, Evans *et al.* 3027 (MO, US); Wilhelmina Mts., at summit of Frederick Top, 3 km SE of Juliana Top, Irwin *et al.* 54963 (NY, US). French Guiana: Piste de St Elie, near Sinnamary, Prévost 4161 (CAY, US); Station des Nouragues, de Granville *et al.* 11132 (B, CAY, NY, P, US).

Phenology: Collected in flower March-November, in fruit January and July-October.

Vernacular names: French Guiana: (from Grenand *et al.*, 2004, p. 381) mâle herbe serpent (Créole), yawayi lo (Wayãpi).

Uses: Used in local medicine against scorpion stings; cultivated as an ornamental, several color-forms have been introduced in cultivation.

10. **Nautilocalyx porphyrotrichus** (Leeuwenb.) Wiehler, Phytologia 27: 308. 1973. – *Episcia porphyrotricha* Leeuwenb., Acta Bot. Neerl. 7: 311, 411, fig. 26. 1958. Type: Guyana, Davenport 7 (holotype K).

Terrestrial herb, ca. 15 cm tall. Stem sappy, decumbent, setose at apex, glabrescent. Leaves unequal in a pair; petiole 0.8-4 cm long, hirsute; blade chartaceous when dry, oblong-lanceolate, larger blade 7-16 x 2.5-4 cm, margin serrate, apex acuminate, base cuneate or decurrent into petiole, above hirsute, below hirsute. Flowers (1-)2, in epedunculate fascicles; pedicel 2-2 cm long, purple-hirsute. Calyx purplish-green, lobes free, 4 erect, subequal, linear, 1-1.5 x 0.1-0.2 cm, dorsal one curved around spur, somewhat smaller, margin with some teeth near apex, apex acuminate, outside purple-hirsute, inside sparsely pubescent; corolla erect in calyx, crimson, 4.7-5.4 cm long, tube trumpet-shaped, 3.8-4.2 cm long, base spurred, 0.3-0.4 cm wide, middle cylindric, hardly ventricose, throat slightly contracted, 0.5-0.6 cm wide, outside villous, inside pubescent with glandular hairs in throat, limb 1.7-2.2 cm wide, lobes subequal, spreading, rounded, 0.6-1 x 0.6-1 cm, margin obscurely repand-serrate; stamens included, inserted on base of corolla; ovary ovoid, 0.4 x 0.2-0.5 cm, hirsute, style 2.8 cm long, glabrous, stigma 2-lobed. Mature fruit not seen.

Distribution: Venezuela (Amazonas, Bolívar) and Guyana; in Guyana in moist places in rainforests, at 640-800 m alt.; > 30 collections studied (GU: 4).

Specimens studied: Guyana: Cuyuni-Mazaruni Region, return from Kaika, Piaitma tipu, McDowell *et al.* 3054 (US); Pakaraima Mts., Wenamu R., Davenport 7 (K); Upper Mazaruni, Kamarang R., Tillett 45749 (NY, US), 45808 (US, NY).

Phenology: Collected in flower in July, September and October.

11. **Nautilocalyx punctatus** Wiehler, Selbyana 5: 40. 1978. Type: Cult. Hort. Selby, Wiehler 77131 (holotype SEL, isotypes BH, K, MO, MY, NY, SEL, U, UMICH, US, VEN), originally from Venezuela, Amazonas, Steyermark 103207.

Terrestrial herb, 10-40 cm tall. Stem sappy, erect, decumbent at base, sericeous. Leaves subequal in a pair; petiole 0.3-11 cm long, sericeous; blade membranous when dry, oblanceolate, 15-25 x 4.5-9.5 cm, margin serrate, apex acuminate, base decurrent, above and below sericeous, purplish below. Flowers 1-8, in epedunculate fascicles; pedicel 1-2.5 cm long, sericeous. Calyx pinkish, lobes free, 4 erect, subequal, lanceolate, 1.9-2.5 x 1.1 cm, dorsal one slightly curved around spur and somewhat smaller, margin serrate, apex acuminate, outside sericeous, inside glabrous; corolla horizontal in calyx, white, ca. 4.5 cm long, tube tubular, 3.5-4 cm long, base spurred, 0.3 cm wide, middle cylindric, throat slightly contracted dorsiventrally, 1 cm wide, outside sericeous, inside with glandular trichomes dorsally, limb 2.5 cm wide, lobes subequal, spreading, rounded, 0.7-1.1 x 1.2-1.3 cm, margin ciliate; stamens included, inserted 0.8 cm from base of corolla; ovary ovoid, 0.4 x 0.3 cm, sericeous, style 3-3.5 cm long, pilose, stigma 2-lobed. Mature capsule ovoid, 0.4-0.5 x 0.3-0.4 cm.

Distribution: Venezuela (Amazonas, Yarucuy) and Guyana; forest on sand, at 160-180 m alt.; 18 collections studied (GU: 3).

Specimens studied: Guyana: Rewa R., near Corona Falls, Jansen-Jacobs *et al.* 5821 (U, US); U. Takutu-U. Essequibo Region, Kuyuwini R., Clarke 4392 (US); Mts. Wassarai, Clarke *et al.* 7998 (US).

Phenology: Collected in flower in September.

17. **PARADRYMONIA** Hanst., Linnaea 26: 207. 1854 ('1853'). – *Episcia* Mart. sect. *Paradrymonia* (Hanst.) Leeuwenb., Acta Bot. Neerl. 7: 311. 1958.
Type: P. glabra (Benth.) Hanst. (Centrosolenia glabra Benth.) [Paradrymonia ciliosa (Mart.) Wiehler]

Episcia sect. *Pagothyra* Leeuwenb., Acta Bot. Neerl. 7: 312. 1958.
Type: E. maculata Hook. f.
Episcia sect. *Salpinganthus* Leeuwenb., Acta Bot. Neerl. 7: 313. 1958.
Type: E. densa C.H. Wright

Terrestrial or epiphytic, caulescent, erect, ascending, or scrambling herbs, subshrubs, or lianas, without modified stems. Stems branched or unbranched. Leaves opposite, often appearing rosulate, equal to strongly unequal in a pair, venation pinnate, foliar nectaries absent. Flowers axillary, in pedunculate or epedunculate, usually congested cymes; bracteoles present; pedicellate. Calyx lobes free; corolla white or yellow with red or purple spots or lines, funnel-form or trumpet shaped; stamens included, filaments not connate or basally connate, anthers coherent, dehiscing by longitudinal slits, thecae parallel or divergent at base; staminode small; disc of 1 or 2 glands; ovary superior, stigma capitate, stomatomorphic, or 2-lobed. Fruit a fleshy, white or colored capsule, loculicidally dehiscent, 2-valved, valves opening slightly.
Chromosome number n=9 (Skog 1984).

Distribution: 36 or more species, in C and northern S America; in or at the edge of rainforest; 7 species in the Guianas.

KEY TO THE SPECIES

1 Leaf base long decurrent .... *4. P. ciliosa*
  Leaf base cuneate, rounded, or slightly cordate 2

2 Terrestrial 3
  Epiphyte 4

3 Stem shorter than the leaves, decumbent, apex erect .... *5. P. densa*
  Stem longer than the leaves, applied to the ground or rocks ....
  .... *3. P. campostyla*

4 Leaves strongly unequal in a pair .... *1. P. anisophylla*
  Leaves equal or subequal in a pair 5

5 Adventitious roots borne along a straight longitudinal line at and between
  nodes (like *Hedera helix*) 6
  Adventitious roots borne only at nodes 7

6 Leaf blade < 6 cm long; flowers solitary .... *2. P. barbata*
  Leaf blade > 7 cm long; flowers in a large inflorescence .... *7. P. maculata*

7   Stem lianescent and appressed to the support, or suffrutescent herb; flowers
    1-3 per axil .... *3. P. campostyla*
    Stem erect; flowers several in an axil .... *6. P. longifolia*

1.   **Paradrymonia anisophylla** Feuillet & L.E. Skog, Brittonia 54: 354.
     2003 ('2002'). Type: Guyana, Cuyuni-Mazaruni Region, Partang R.,
     top of Merume Mt., Tillett, Tillett & Boyan 43948 (holotype US,
     isotype NY).   – Fig. 22 A-D

Epiphytic suffrutex, reaching 50 cm long. Stem sappy, pendent, hirsute.
Leaves usually strongly unequal in a pair; petiole 0-1 cm long, hirsute;
blade chartaceous when dry, oblanceolate, blade of large leaf in a pair 3-
7.5 x 0.8-2.0 cm, margin crenate or serrate, apex blunt to acuminate, base
attenuate, somewhat asymmetric, above velvety, below with shorter
hairs. Flowers solitary; epedunculate; pedicel 1.7-2 cm long, hirsute.
Calyx reddish, lobes free, 4 erect, subequal, lanceolate, 1-1.8 x 0.2-0.5 cm,
dorsal one curved around spur, margin entire or 3-5-toothed, apex acute
or acuminate, outside and inside hirsute; corolla oblique in calyx,
creamy, yellowish at throat, 4.5-5 cm long, tube 3.5-4 cm long, base
spurred, 0.5-1 cm wide, middle cylindric, throat 1-2 cm wide, outside
appressed-tomentose, inside puberulent, limb 2-3.5 cm wide, lobes
spreading, suborbicular, 0.5-1 x 0.7-1 cm, ventral lobe larger with
margin fimbriate; stamens included, inserted near base of corolla; ovary
ovoid, 0.2 x 0.15 cm, hirsute, style 1.5 cm long, glabrous, stigma
capitate. Mature capsule subglobose, 0.5 x 0.4 cm.

Distribution: Known from a few localities in Guyana, Mt.
Ayanganna, Wokomung and Merume Mt.; epiphyte in trees in moist
forest on plateaus or mountain tops, at 1070-1160 m alt.; 7 collections
studied (GU: 7).

Specimens studied: Cuyuni-Mazaruni Region, Partang R., top of
Merume Mt., Tillett *et al.* 43948 (NY, US); Potaro-Siparuni Region, Mt.
Ayanganna, Clarke 8950 (US), 9006 (US); 9202 (US); 9618 (US); 9619
(US); Mt. Wokomung, Boom & Samuels 8990 (NY).

Phenology: Flowering observed in June and July, fruiting observed
in June.

2.   **Paradrymonia barbata** Feuillet & L.E. Skog, Brittonia 54: 356.
     2003 ('2002'). Type: Guyana, Cuyuni-Mazaruni Region, near Eping
     R., McDowell & Stobey 3810 (holotype BRG, isotypes K, US).
                                                            – Fig. 23

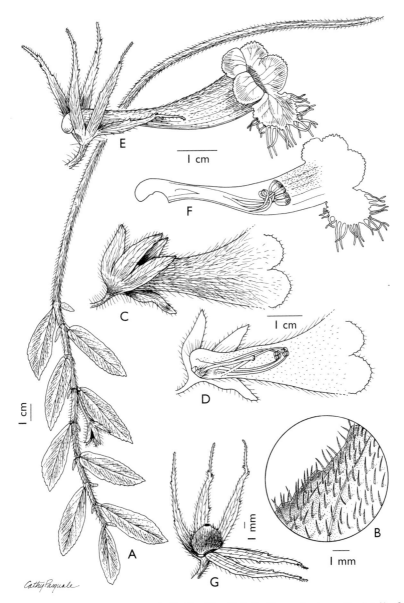

Fig. 22. *Paradrymonia anisophylla* Feuillet & L.E. Skog: A, stem; B, detail of leaf margin; C, flower (limb not attached to corolla tube, ventral lobe not seen); D, stamens, nectary, and pistil. *Paradrymonia ciliosa* (Mart.) Wiehler: E, flower; F, corolla opened to show stamens; G, fruit in persistent calyx. (A-D, Tillett *et al.* 43948; E-F, Grenand *et al.* 2003; G, de Granville 962).

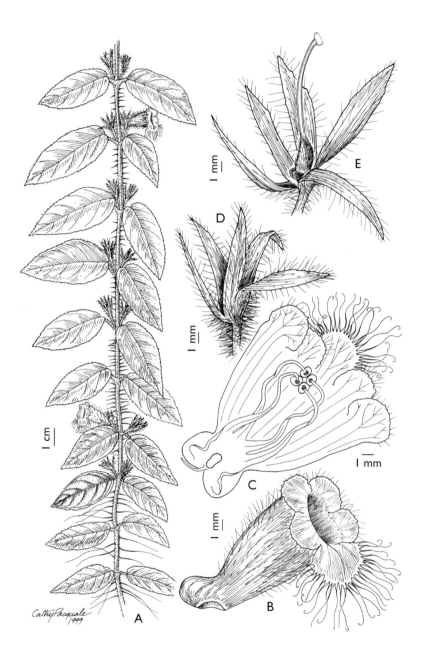

Fig. 23. *Paradrymonia barbata* Feuillet & L.E. Skog: A, climbing stem with adventitious roots; B, corolla; C, corolla opened dorsally to show stamens and staminode; D, calyx; E, nectary gland and pistil. (A-E, McDowell *et al.* 3810).

Epiphytic herb, 30 cm long. Stem sappy becoming sarmentose, creeping, repent or pendent, hirsute. Leaves equal in a pair; petiole 0.5-1 cm long, hirsute; blade chartaceous when dry, lanceolate, 4-5 x 2-2.5 cm, margin crenate, apex acute, base slightly cordate to rounded, above appressed-pubescent, below appressed-pubescent along veins. Flowers solitary; epedunculate; pedicel 0.5-1 cm long, pilose. Calyx lobes free, erect, subequal, narrowly lanceolate, 0.7-1 x 0.1-0.2 cm, margin with 1-4 obscure teeth, apex acute, outside hirsute, inside strigose; corolla oblique in calyx, cream, 1.2-1.5 cm long, tube slightly infundibuliform, 1-1.3 cm long, base spurred, 0.3 cm wide, middle slightly contracted, throat round, 0.5-0.6 cm wide, outside hirsute, inside glabrous, limb 0.8-1.2 cm wide, lobes unequal, spreading, rounded, 0.2-0.4 x 0.15-0.4 cm, margin ventral lobe long-fimbriate; stamens slightly exserted, inserted near base of corolla; ovary oblong, 0.3 x 0.25 cm, appressed-pubescent, style 0.7-0.9 cm long, glabrous, stigma capitate. Mature fruit not seen.

Distribution: Endemic to Guyana from the Cuyuni-Mazaruni Region (near Eping R.); in lowland forest, climbing, radicant on trees, at 120 m alt.; know only from the type collection (GU: 1).

Specimen studied: Guyana: Cuyuni-Mazaruni Region, near Eping R., McDowell & Stobey 3810 (BRG, K, US).

Phenology: Collected in flower in February.

3. **Paradrymonia campostyla** (Leeuwenb.) Wiehler, Selbyana 5: 49. 1978. – *Drymonia campostyla* Leeuwenb., Acta Bot. Neerl. 7: 305, 393, fig. 19, 20. 1958. Type: Suriname, Krammaman, near Kadjoe, Jonker & Jonker 625 (holotype U, isotype US).

Terrestrial creeping or epiphytic plants, rooted at nodes, ultimately developing into suffrutescent herb (like *Drymonia coccinea* or *Hedera helix*). Stem sappy, creeping, hirsute, especially near apex. Leaves equal in a pair; petiole 0.6-4.5 cm long, hirsute; blade chartaceous when dry, often obliquely oblong-ovate or oblong-elliptic, 2.5-10 x 1.2-4.5 cm, margin subentire or obscurely repand-serrate, apex acute or acuminate, base cuneate or rounded, above hirsute or pilose, below hirsute or pilose, especially on veins. Flowers 1-3, in fascicles; peduncle very short or obsolete; pedicel 0.5-2 cm long, hirsute. Calyx green, lobes free, oblique, 4 subequal, ovate-lanceolate, 2-3.5 x 0.5-1.1 cm, dorsal one about half as long, margin repand-serrate, apex long-acuminate, outside hirsute, inside hirsute; corolla oblique in calyx, white, 3.5-6 cm long, tube 3-4.5 cm long, base spurred, 0.3-0.5 cm wide, middle not or hardly ventricose, throat slightly contracted, 1-1.5 cm wide, outside hirsute, inside partially pubescent with glandular hairs, minutely verrucose at throat, limb 2-3.5 cm wide, lobes subequal, spreading, suborbicular, 0.6-1.3 x 0.6-1.3 cm, margin entire; stamens included; ovary ovoid, 0.4-0.5 x 0.25-0.3 cm, hirsute, style 2.5-4 cm long, shortly hirsute, stigma 2-lobed. Mature capsule globose, ca. 1 cm in diameter.

Distribution: Suriname, French Guiana, and Brazil (Amapá); in rainforest, at 20-670 m alt.; 60 collections studied (SU: 15; FG: 50).

Selected specimens: Suriname: Krammaman, near Kadjoe, Jonker & Jonker 625 (U, US); near Brownsberg, Wessels Boer 634 (U, US), Koster LBB 13025 (BBS, US). French Guiana: Mt. Galbao, de Granville *et al.* 8733 (CAY, NY, US); Pic Coudreau, de Granville *et al.* 11837 (CAY, NY, P, U, US); Papaïchton, Sastre & Bell 8140 (CAY, P, US).

Phenology: Collected in flower February-September and December, in fruit March, April, July.

Vernacular names: French Guiana: yamuleka asili (Wayãpi; de Granville 2449 and Grenand *et al.*, 1987), yamuleka'a sili (Wayãpi; Grenand *et al.*, 2004); ewoi asikaluwu (Wayãpi; Prévost & Grenand 1988).

Use: External febrifuge (Wayãpi).

4. **Paradrymonia ciliosa** (Mart.) Wiehler, Phytologia 27: 308. 1973. – *Hypocyrta ciliosa* Mart., Nov. Gen. Sp. Pl. 3: 53. 1829. – *Episcia ciliosa* (Mart.) Hanst. in Mart., Fl. Bras. 8(1): 403. 1864. – *Columnea ciliosa* (Mart.) Kuntze, Revis. Gen. Pl. 2: 472. 1891. Type: Brazil, Amazonas, Martius 3117 (holotype M). – Fig. 22 E-G

Centrosolenia glabra Benth., Bot. Mag. 76: ad pl. 4552. 1850. – *Paradrymonia glabra* (Benth.) Hanst., Linnaea 26: 207. 1854 ('1853'). – *Episcia glabra* (Benth.) Hanst., Linnaea 34: 349. 1865 ('1865-1866'). Type: Cult. Hort. Kew. (holotype K), originally from Venezuela, Distríto Federál.

Terrestrial or epiphytic herb, about 30-40 cm tall. Stem sappy becoming subwoody at base, procumbent, hirsute, glabrescent towards base. Leaves strongly unequal in a pair; petiole 0.6-1.4 cm long, hirsute; blade chartaceous when dry, oblong-lanceolate, larger blade 17-40 x 4.5-11 cm, margin repand-serrate, ciliate, apex acuminate, base long-decurrent into petiole, above hirsute, below hirsute, especially on midrib and veins.

Flowers many, in fascicles; peduncle very short or obsolete; pedicel 1-2.5 cm long, hirsute. Calyx purplish or crimson, lobes free, 4 erect, dorsal lobe curved around spur, subequal, linear, 1-2 x 0.1-0.2 cm, margin with some teeth, apex long-acuminate, outside hirsute, inside glabrous; corolla slightly oblique in calyx, white or creamy, 3-4 cm long, tube infundibuliform, about 2-3.3 cm long, base shortly spurred, 0.15 cm wide, middle gradually widened, not or hardly ventricose, throat somewhat contracted, 0.5-0.8 cm wide, outside pubescent, inside glabrous, limb 0.8-1.5 cm wide, lobes unequal, spreading, broadly rounded, 0.3-0.7 x 0.3-0.4 cm wide, margin fimbriate; stamens included, inserted on base of corolla; ovary ovoid, 0.3-0.4 x 0.2-0.25 cm, softly hirsute, style 1.5-3 cm long, sparsely pubescent, stigma capitate. Mature capsule included in persistent calyx, 1 cm long, 0.8 cm wide.

Distribution: Colombia, Ecuador, Peru, Venezuela, the Guianas, and Brazil (Amazonas); in rainforest, at 80-580(-1000) m alt.; > 100 collections studied (GU: 6; SU: 2; FG: 6).

Selected specimens: Guyana: Kaieteur Plateau, Cowan & Soderstrom 2208 (NY, US); Cuyuni-Mazaruni Region, W branch of Eping R., McDowell & Stobey 3917 (NY, U, US). Suriname: Mts. Tumuc-Humac, Cr. Waamahpann, de Granville 962 (CAY); Haut Litany, de Granville et al. 12067 (CAY, US). French Guiana: Armontabo Cr., Grand Saut, Prévost 1903 (CAY, P, US); Route Régina-Saint Georges D.Z. 5-P.K. 43, Cremers et al. 12030 (B, CAY, MO, NY, P, U, US).

Phenology: Collected in flower February-June and August, in fruit in August.

5. **Paradrymonia densa** (C.H. Wright) Wiehler, Selbyana 5: 50. 1978. – *Episcia densa* C.H. Wright, Bull. Misc. Inform. Kew 1895: 17. 1895. – *Centrosolenia densa* (C.H. Wright) Sprague, Bull. Misc. Inform. Kew 1912: 87. 1912. Type: Guyana, Masouria R., Jenman 2414 (holotype K).

Terrestrial herb, 30-60 cm high. Stem fleshy, ascending, glabrous, sometimes puberulous at apex. Leaves equal or subequal in a pair; petiole (1-)6-12(-14) cm long, glabrous; blade chartaceous when dry, oblong-elliptic or oblong-lanceolate, 5-30 x 1-12 cm, margin subentire to crenate-serrate, apex acute, acuminate, or sometimes obtuse, base cuneate, rounded, or sometimes obliquely subcordate, above mostly glabrous or nearly so, sometimes strigose to subtomentose, below mostly glabrous or nearly so, sometimes sparsely pilose. Flowers mostly numerous and aggregated, umbellate or cymose; peduncle up to 1 cm long, minutely puberulous or glabrous; pedicel 0.7-3 cm long, practically glabrous. Calyx purple, lobes 4, erect, connate $^1/_3$ their length, subequal, obliquely oblong-lanceolate, 1.2-3.3 x 0.3-0.8 cm, dorsal one mostly free,and slightly smaller, margin repand-serrate, apex acute, outside sparsely pilose or glabrous, inside sparsely pilose or glabrous; corolla oblique or horizontal in calyx, white or pale stramineous, sometimes with purple or purple-tipped lobes, (3-)4-6 cm long, tube trumpet-shaped, 2.5-4.7 cm long, base shortly spurred, 0.3-0.4 cm wide, middle not or hardly ventricose, slightly bent downwards, throat not or hardly contracted, 0.5-1.2 cm wide, outside hirsute, inside glabrous, limb 1.5-1.8 cm wide, lobes subequal, spreading, ventral lobe somewhat concave, suborbicular, 0.5-0.7 x 0.5-0.7 cm, margin entire; stamens included, inserted at base of corolla; ovary ovoid, 0.2-0.4 x 0.1-0.2 cm wide, hirsute, style 2-4 cm long, minutely puberulous or glabrous, stigma capitate. Mature capsule globose or nearly so, ca. 1 cm in diameter.

Distribution: Endemic to the Guianas; on forest floor, often in clearings, along the rivers, often on white sand or lateritic soil, at 0-700 m alt.; > 100 collections studied (GU: 24; SU: 1; FG: 50).

Selected specimens: Guyana: West Demerara Region, Mabura Hill area, Pipoly *et al.* 7572 (BRG, NY, U, US), Maas *et al.* 5869 (BBS, COL, MO, P, U, US). Suriname: Wilhelmina Mts., Stahel 404 (U, US). French Guiana: near Régina, Bordenave 1339 (CAY, P, US); Bassin de l'Approuague, Savane Roche de Virginie, Cremers *et al.* 11771 (B, CAY, MO, NY, P, U, US).

Phenology: Collected in flower throughout the year, in fruit April-June, September, and October.

6. **Paradrymonia longifolia** (Poepp.) Wiehler, Selbyana 5: 54. 1978. – *Drymonia longifolia* Poepp. in Poepp. & Endl., Nov. Gen. Sp. Pl. 3: 4. 1840. – *Episcia longifolia* (Poepp.) Hanst., Linnaea 34: 347. 1865 ('1865-1866'). Type: Peru, Huánuco, Poeppig 1671 (holotype W, isotypes W(2)).

Epiphytic herb, 0.5-1 m tall. Stem slightly fleshy, ascending, villous. Leaves subequal in a pair; petiole 1-4 cm long, villous; blade chartaceous when dry, elliptic or oblong-elliptic, 13-23 x 4-6 cm, margin serrate, apex slightly acute, base cuneate to decurrent, above villous, below villous. Flowers numerous, fasciculate; pedunculate; pedicel 0.4-1 cm long, villous. Calyx cream-white to yellow, lobes free, erect, subequal, oblong, 2-4 x 0.2-0.5 cm, margin entire to wavy, apex acute to obtuse, outside villous, inside appressed-pubescent; corolla slightly oblique in calyx, greenish white to yellow, 2-2.5 cm long, tube funnelform, 1.5-2 cm long, base shortly spurred, 0.3-0.4 cm wide, middle cylindric, throat 0.8-1 cm wide, outside appressed-pubescent, inside glabrous except ventrally, limb 1.5-1.8 cm wide, lobes subequal, spreading, suborbicular, 0.3-0.6 x 0.6-0.9 cm, margin entire; stamens included; ovary ovoid, 0.4 cm long, 0.25 cm wide, densely hairy, style 1.0-1.2 cm long, stigma stomatomorphic. Mature capsule yellowish, ovoid, laterally compressed, 0.7 x 0.8 cm.

Distribution: Colombia, Ecuador, Peru, French Guiana; rainforest, climbing on trunks, at 350-400 m alt.; 20 collections studied (FG: 1).

Specimen studied: French Guiana: Mts. de Kaw, de Granville 2933 (CAY).

Phenology: Collected in bud in June.

Note: Description of flowers and fruits from collections from Ecuador, Peru and Colombia.

7. **Paradrymonia maculata** (Hook. f.) Wiehler, Selbyana 5: 57. 1978. – *Episcia maculata* Hook. f., Bot. Mag. 116: ad pl. 7131. 1890. Type: Cult. Hort. Kew., 2 Sep 1859 (holotype K), originally from Guyana.

Epiphytic vine on tree trunks, up to 2 m tall. Stem sappy, creeping or climbing, hirsutulous at apex, glabrescent. Leaves equal or subequal in a pair; petiole (1.5-)5-10(-19) cm long, sparsely pilose to glabrous; blade papyraceous when dry, elliptic or oblong-elliptic, 7.5-29 x 3.5-12 cm, margin serrate with more or less pronounced teeth, apex acuminate, base cuneate, rounded, or occasionally subcordate, above sparsely pilose to glabrous, below sparsely pubescent, especially on midrib and veins. Flowers numerous, racemose; peduncle 1-3 cm long, sparsely pubescent; pedicel 0.5-2.2 cm long, sparsely pubescent. Calyx colored like bracts, lobes nearly free, 4 spreading, subequal, leafy, linear-lanceolate, narrowed towards base, 2-4 x 0.4-0.7 cm, dorsal lobe curved around spur, half as large as others, margin serrate towards apex, apex acute or acuminate, outside and inside sparsely appressed-pubescent; corolla oblique in calyx, yellow or pale creamy, spotted with bright red, brown, or purple-brown, 4.4-5.5 cm long, tube infundibuliform, 2.7-3.5 cm long, base spurred, 0.4-0.5 cm wide, middle widened towards throat, not ventricose, throat not contracted, 1-1.8 cm wide, outside with minute hairs, inside glabrous, limb 3-4 cm wide, lobes subequal, 4 spreading, the ventral one tightly inflexed and closing the throat, suborbicular, 1-1.2 x 1-1.2 cm, margin entire; stamens included, inserted slightly under middle of corolla tube; ovary ovoid, 0.4-0.5 x 0.3-0.4 cm wide, pubescent, style 2.2-3 cm long, glabrous, stigma obscurely 2-lobed. Mature capsule subglobose, 1-1.5 x 0.8-1.4 cm.

Distribution: Endemic to the Guianas and Delta Amacuro, Venezuela; in rainforest, at 0-100 m alt.; 35 collections studied (GU: 21; FG: 6).

Selected specimens: Guyana: Barima-Waini Region, Barima R., 15 miles E of Arakaka, Pipoly *et al.* 8059 (BRG, CAY, US); U. Takutu-U. Essequibo Region, near Dadanawa, de la Cruz 1535 (CM, F, US). French Guiana: Lower Oyapock basin, Cr. Armontabo, Prévost & Grenand 1996 (CAY, P), Cr. Gabaret, Cremers 9951 (CAY, NY, P, U, US).

Phenology: Collected in flower in January, April, June, August, September, and December, in fruit in January, June, and December.

Note: The plant apparently germinates on the forest floor, climbs lower tree trunks and slowly decays from bottom up.

18. **RHOOGETON** Leeuwenb., Acta Bot. Neerl. 7: 321. 1958.
   Type: R. cyclophyllus Leeuwenb.

Terrestrial acaulescent erect herbs, sometimes tuberous. Stems unbranched. Leaves radical, rosulate, equal to strongly unequal in a pair, venation pinnate, stomata randomly scattered, foliar nectaries absent. Flowers axillary, 1-6, in umbellate, subcymose, or thyrsoid, inflorescences; long pedunculate; bracteoles small, lanceolate; pedicellate. Calyx lobes free; corolla orange or red, nearly trumpet-shaped, but slightly bent downwards; stamens included, filaments not connate, anthers all coherent, coherent in 2 pairs, or all free, dehiscing by longitudinal slits, thecae parallel; staminode very small; disc a single large dorsal gland; ovary superior, stigma 2-lobed. Fruit a dry, loculicidally dehiscent, 2-valved capsule, valves opening to 180°. Chromosome number unknown.

Distribution: A genus of 2 species of the Guayana Highlands of Venezuela and Guyana, and neighbouring Brazil; growing on wet rocks in montane forest, often near waterfalls.

KEY TO THE SPECIES

1 Leaf blade nearly orbicular, rounded at apex, cordate at base, crenate-serrate, without bulbil .... *1. R. cyclophyllus*
  Leaf blade ovate or oblong-ovate, apex acute to acuminate, base rounded or cuneate, biserrate-dentate, occasionally producing marginal bulbils .....
  .... *2. R. viviparus*

1. **Rhoogeton cyclophyllus** Leeuwenb., Acta Bot. Neerl. 7: 322, 429, fig. 31. 1958. Type: Guyana, Cuyuni-Mazaruni Region, Pakaraima Mts., Mt. Ayanganna, Maguire *et al.* 40585 (holotype NY).
  – Fig. 24 E-G

Terrestrial herb, 5-15 cm tall (including leaves). Stem succulent, ascending, villous. Leaves subequal in a pair; petiole 0.5-3.0 cm long, sparsely pubescent; blade papyraceous when dry, nearly orbicular, 1.5-4.5 x 1.5-3.5 cm, margin crenate-serrate, apex rounded, base cordate, above pubescent, below with a few scattered hairs. Flowers 1-6, subcymose or thyrsoid; peduncle 3-15 cm long, sparsely pubescent; pedicel 0.5-1 cm long, pubescent. Calyx subcampanulate, green, lobes free, 4, subequal, dorsal lobe somewhat smaller and narrower, curved around spur, all oblong-lanceolate, 0.25-0.4 x 0.1-0.15 cm, margin entire, apex acute to acuminate, outside and sparsely pubescent; corolla oblique in calyx, orange, 2-2.2 cm long, tube trumpet-shaped, 1.4-1.6 cm long, base spurred, 0.2 cm wide, middle ventricose, throat slightly contracted, 0.5 cm wide, outside and inside sparsely pubescent, limb 1.1-1.7 cm wide, lobes unequal, spreading, obovate to rounded, 0.35-0.9 x 0.35-0.7 cm, margin subcrenate-serrate; stamens included, inserted near base of corolla; ovary ovoid, 0.2 x 0.15 cm, glabrous, style ca. 1 cm long, glabrous, stigma 2-lobed. Mature fruit not seen.

Distribution: Endemic to Guyana; on montane slopes on dripping rocks, at 200-1600 m alt.; 8 collections studied (GU:8).

Selected specimens: Guyana: Cuyuni-Mazaruni Region: Headwaters of Kangu R., 4 km NW of E peak of Mt. Ayanganna, Pipoly *et al.* 11056 (NY); upper slopes of Mt. Wokomung, Boom & Samuels 9156 (US); Potaro-Siparuni Region: Kaieteur Plateau, bottom of Potaro Gorge near Kaieteur Falls, Cowan & Soderstrom 2149 (E, NY, US); Mt. Kopinang, Hahn *et al.* 4346 (US); Kaieteur, Potaro R., Jenman 896 (BRG, K).

Phenology: Collected in flower in March, April, and July, September, and October.

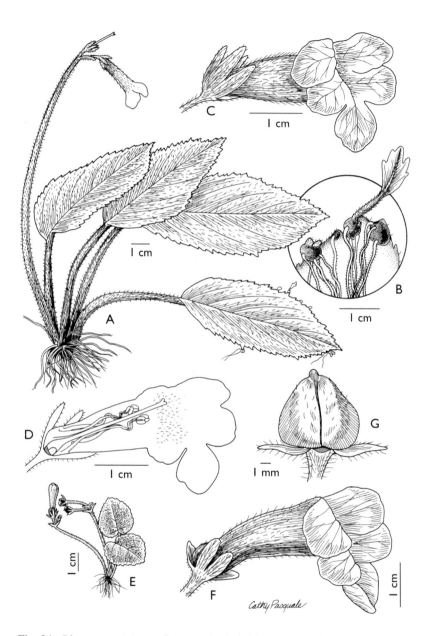

Fig. 24. *Rhoogeton viviparus* Leeuwenb.: A, habit; B, detail of lower leaf margin showing bulbils; C, flower; D, flower opened to show stamens, nectary and pistil. *Rhoogeton cyclophyllus* Leeuwenb.: E, habit; F, flower; G, unripe capsule. (A-B, Kvist *et al.* 370; C-D, Maguire *et al.* 46051A; E-F, Cowan & Soderstrom 2149; G, Hahn *et al.* 4657).

2. **Rhoogeton viviparus** Leeuwenb., Acta Bot. Neerl. 7: 323, 431, fig. 32. 1958. Type: Guyana, Potaro-Siparuni Region, Kaieteur, Jenman 880 (holotype P, isotypes BRG, K, U). – Fig. 24 A-D

Rhoogeton leeuwenbergianus C.V. Morton, Bol. Soc. Venez. Ci. Nat. 23: 80. 1962. Type: Venezuela, Bolívar, Steyermark & Nilsson 24 (holotype US, isotypes F, NY, U).

Terrestrial herb, 10-30 cm tall (including leaves). Stem succulent, ascending, villous. Leaves subequal in a pair; petiole 0.7-18 cm long, strigillose; blade papyraceous when dry, ovate or oblong-ovate, 3.5-18.5 x 2-8 cm, margin biserrate-dentate, occasionally producing marginal bulbils, apex acute to acuminate, base rounded or cuneate, above sparsely strigillose to glabrous, below strigillose, especially on veins. Flowers 1-6, umbellate or subcymose; peduncle 7-38 cm long, sparsely strigillose at base, glabrescent above; pedicel 0.5-2 cm long, strigillose. Calyx subcampanulate, green to reddish, lobes free, 4 lobes subequal, dorsal lobe smaller and curved around spur, all oblong, 0.6-1 x 0.15-0.3 cm, margin entire or sparsely toothed, apex acute to acuminate, outside sparsely strigillose, inside sparsely strigillose; corolla oblique in calyx, red-scarlet to orange-rose outside, 2.5-3.7 cm long, tube trumpet-shaped, 1.8-2.8 cm long, base spurred, 0.2-0.3 cm wide, middle slightly ventricose, throat slightly contracted, 0.5-0.7 cm wide, outside sparsely strigillose, inside sparsely pilose towards mouth, limb 1.2-2.5 cm wide, lobes subequal, spreading to patent, rounded, 0.4-1 x 0.4-1 cm, margin subcrenate-serrate; stamens included, adnate for 0.1-0.2 cm to base of corolla tube; ovary ovoid, 0.3-0.4 x 0.2-0.25 cm, pubescent, style ca. 2 cm long, sparsely pubescent, stigma 2-lobed. Mature capsule not seen.

Distribution: Eastern Venezuela (Bolívar), Guyana, and Brazil; growing on wet rocks near rivers and waterfalls, at 180-2000 m alt.; 23 collections studied (GU: 15).

Selected specimens: Guyana: Cuyuni-Mazaruni Region: Karowtipu Mt., Boom *et al.* 7734 (NY, P, U, US); Mt. Roraima Escarpment, R. Persaud 80 (BRG, K); Upper Mazaruni R. basin, Mt. Ayanganna, Tillett & Tillett 45186 (NY); Kamarang R., Tillett & Tillett 45563 (NY, US); Potaro-Siparuni Region: Mt. Ayanganna, Clarke 9255 (U, US); Kaieteur Plateau, bottom of Potaro Gorge near Kaieteur Falls, Cowan & Soderstrom 2148 (E, F, MO, NY, US); beneath Kaieteur Falls, Kvist *et al.* 370 (AAU, B, BRG, NY, P, U, US); Pakaraima Mts., Kopinang Falls, Maguire *et al.* 45984A (B, GH, K, NY, U, US).

Phenology: Collected in flower February-April, August-October.

19. **SINNINGIA** Nees, Ann. Sci. Nat. (Paris) 6: 297. 1825.
    Type: S. helleri Nees

Terrestrial or epipetric, rarely epiphytic caulescent decumbent to erect herbs or subshrubs, tuberous. Stems branched or unbranched. Leaves opposite or in whorls, or congested on short stems, nearly equal in a pair or whorl, venation pinnate, foliar nectaries absent. Flowers axillary, 1 to many, umbellate, cymose, paniculate, or appearing terminal in racemes; usually epedunculate (in Guianan species); bracteoles present; pedicellate. Calyx lobes connate at base; corolla usually red or orange, rarely yellow, purplish or white, campanulate to cylindric; stamens exserted, filaments not connate, anthers coherent, dehiscing by longitudinal slits, thecae parallel to divergent; staminode small; disc of 1-5 glands, sometimes with 2 larger and connate; ovary half-inferior to almost superior, stigma stomatomorphic to capitate. Fruit a dry, brown, loculicidally dehiscent, 2-valved capsule, valves opening slightly. Chromosome number n=13 (Skog 1984).

Distribution: A wide-ranging genus of about 60-65 species from Vera Cruz in Mexico south to northern Argentina, Bolivia, Uruguay, and with many species in Brazil; growing on granitic rocks, or in savannas; 2 species occur in the Guianas, including the widespread *S. incarnata*.

Note: The best-known species in the genus is *Sinningia speciosa* (Lodd.) Hiern and its cultivars, known as the 'Florist's Gloxinia'. In the Gesneriaceae this species is second only to the 'African Violet' (*Saintpaulia* spp.) in popularity for growing as pot plants, and has been in cultivation since 1817.

LITERATURE

Chautems, A. 1990. Taxonomic revision of Sinningia Nees: Nomenclatural changes and new synonymy. Candollea 45: 381-388.
Chautems, A. 1991. Taxonomic revision of Sinningia Nees (Gesneriaceae), 2: New species from Brazil. Candollea 46: 411-425.
Chautems, A. 1995. Taxonomic revision of Sinningia Nees (Gesneriaceae), 3: New species from Brazil and new combinations. Gesneriana 1: 8-14.
Chautems, A. & A. Weber. 1999. Shoot and inflorescence architecture of the neotropical genus Sinningia (Gesneriaceae). In M.H. Kurmann & A.R. Jemsley, The Evolution of Plant Architecture. pp. 305-322.

Moore, H.E. 1954. A proposal for the conservation of the name Rechsteineria. Baileya 2: 24-29.

Moore, H.E. 1973. Comments on cultivated Gesneriaceae. Baileya 19: 35-41.

KEY TO THE SPECIES

1 Calyx lobes acute or acuminate, less than ¹/₂ as long to equalling the tube; leaves tomentose above . . . . *1. S. incarnata*
Calyx lobes long-acuminate, 1-2 times longer than the tube; leaves villose or pilose above . . . . *2. S. schomburgkiana*

1. **Sinningia incarnata** (Aubl.) D.L. Denham, Baileya 19: 126. 1974.
    – *Besleria incarnata* Aubl., Hist. Pl. Guiane 2: 635, 4: pl. 256. 1775.
    – *Fimbrolina incarnata* (Aubl.) Raf., Sylva Tellur. 71. 1838. –
    Rechsteineria incarnata (Aubl.) Leeuwenb., Acta Bot. Neerl. 7: 320,
    425. 1858. Type: French Guiana, basin of Galibi Cr., Aublet s.n.
    (holotype BM).

    Gesnera aurantiaca Hanst., Ind. Sem. Hort. Berol., App. 1861: 8. 1861, as
    'Gesnera aurantiaca'. – *Rechsteineria aurantiaca* (Hanst.) Kuntze, Revis.
    Gen. Pl. 2: 474. 1891. Type: Cult. Hort. Berol. (holotype B destroyed),
    originally from Venezuela, collected by Gollmer.
    Rechsteineria faucidens Hoehne var. *parvifolia* Hoehne, Sellowia 9: 75.
    1958. Type: Brazil, Amapá, Froes 25820 (holotype, IAN).

Terrestrial herb or subshrub, to 1 m tall. Stem sappy becoming subwoody, erect, pubescent to villous, especially towards apex. Leaves equal or subequal in a pair; petiole 0.1-4 cm long, villous; blade papyraceous when dry, oblong, obovate, or elliptic, 2.5-9.5 x 1-4 cm, margin crenate, apex acute, base attenuate into petiole or acute, above tomentose, below tomentose. Flowers 1-5, thyrsoid; peduncle very short or lacking, 0-0.2 cm long, pubescent; pedicel 0.5-4 cm long, pubescent to villous. Calyx campanulate, green or reddish, lobes connate at base, tube 0.3-1 cm, free portion of lobes erect, subequal, broadly triangular, 0.4-0.7 x 0.2-0.6 cm, margin entire, apex acute to acuminate, outside and inside pubescent to pilose and glandular; corolla erect in calyx, red or red-orange outside, 2.7-4.6 cm long, tube cylindric, 3-4 cm long, base dorsally gibbous, ca. 0.7 cm wide, middle broader, throat narrowed just below limb, 0.5-0.8 cm wide, outside pubescent to hirsute, inside glabrous, limb 1.2-2.0 cm wide, lobes unequal, upper lobes connate into

a hood, erect, ca. 1 x 1 cm, margin entire, lateral lobes erect, rounded to
truncate, 0.1-0.3 x 0.3-0.6 cm, margin entire, basal lobe erect, broadly
rounded, ca. 0.2 x 0.4-0.6 cm, margin entire; stamens exserted, inserted
at base of corolla tube; ovary conic, ca. 0.4 x 0.3 mm, pubescent, style
ca. 4 cm long, pubescent to pilose, stigma stomatomorphic. Mature
capsule brownish, ovoid, ca. 1 x 0.4-0.7 cm.

Distribution: Mexico to northern Brazil, Colombia, Venezuela and the
Guianas; found in open areas in forests, or on steep slopes or granitic
outcrops, at 0-670 m alt.; > 300 specimens studied (GU: 13; SU: 8; FG: 12).

Selected specimens: Guyana: Demerara-Mahaica Region, E
Demerara, Parker s.n. (K); Pomeroon-Supenaam Region, W Tapakuma
Lake dam, Hoffman et al. 2835 (U, US); Upper Takutu-Upper Essequibo
Region, Towatawan Mt., Gillespie 1973 (AAU, B, BRG, CAY, NY, U,
US); Sand Cr., Wilson-Browne 70 (K, NY). Suriname: Nickerie Distr.,
Morro Grande, Sipaliwini savanna, Oldenburger et al. 866 (BBS, NY, U);
Saramacca Distr., Voltzberg, Schulz LBB 10601 (BBS, U); Upper
Saramacca R., Pulle 492 (U, US). French Guiana: Mts. Tumuc-Humac, de
Granville 11739 (B, CAY, K, NY, U, US); Roche No. 1, Akouba Booa goo
Soula, Bassin du Haut-Marouini, de Granville et al. 9780 (CAY, US);
Fleuve Approuague, au saut Grand Canori, Oldeman B-1984 (CAY, P, U).

Phenology: Collected in flower from February to November.

Vernacular name: French Guiana: yawalemo (Wayampi; Sastre
4720).

Use: Parker s.n. (K) reports "the plant was used as a cephalic snuff".

2. **Sinningia schomburgkiana** (Kunth & Bouché) Chautems,
Candollea 45: 386. 1990. – *Gesneria schomburgkiana* Kunth &
Bouché, Ind. Sem. Hort. Bot. Berol. 1844: [10]. 1844. –
Rechsteineria schomburgkiana (Kunth & Bouché) Kuntze, Revis.
Gen. Pl. 2: 474. 1891. Type: Guyana, Kanuku Mts., Ro. Schomburgk
118.S (holotype B destroyed, lectotype K, isolectotype BM)
(designated by Chautems 1990: 386). – Fig. 25

Gesneria guianensis Benth., London J. Bot. 5: 360. 1846. Type: Guyana,
Kanuku Mts., Ro. Schomburgk 118.S (holotype K, isotype BM).
Rechsteineria crenata Fritsch in Pilg., Notizbl. Bot. Gart. Berlin-Dahlem 6:
381. 1915. Type: Brazil, Rio Branco, Ule 8320 (holotype B destroyed,
lectotype K, here designated, isolectotypes F, G, L, UC).

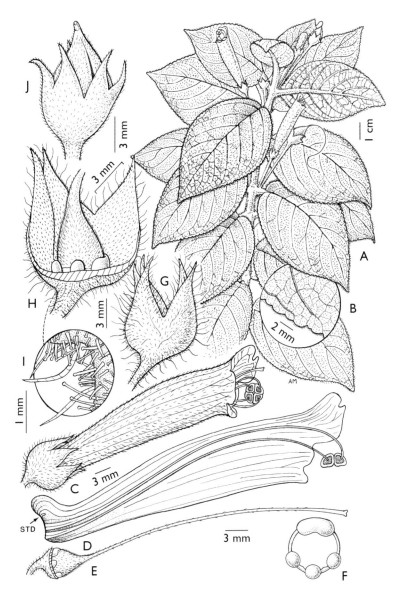

Fig. 25. *Sinningia schomburgkiana* (Kunth & Bouché) Chautems: A, flowering stem; B, detail of leaf margin from below; C, flower; D, corolla opened to show stamens and staminode [STD]; E, nectary glands and pistil; F, nectary glands from above; G, calyx; H, flower with corolla and stamens removed, to show ovary and nectary glands; I, detail of calyx base showing glandular and eglandular hairs; J, capsule. (A, E, H-J, Harrison 1346; B, F-G, A.C. Smith 3652; C-D, Goodland & Persaud 704).

Terrestrial herb or subshrub, to 1 m tall. Stem sappy becoming subwoody, erect, villous. Leaves equal or subequal in a pair; petiole 0.3-3 cm long, villous; blade papyraceous when dry, oblong or ovate, 4.5-14 x 2.9-7 cm, margin crenate, apex acute, base cuneate or rounded, above villous to pilose, below villous to pilose. Flowers 1-2; peduncle very short or lacking, 0-0.2 cm long, pubescent; pedicel 0.8-3.2 cm long, villous. Calyx campanulate, green, lobes connate at base, tube 0.1-0.3 cm, free portion of lobes erect, subequal, triangular, 0.5-0.9 x 0.2-0.4 cm, margin entire, apex long acuminate, outside villous with glandular and eglandular hairs, inside pubescent; corolla erect in calyx, red, 4-5 cm long, tube cylindric, 3.1-3.6 cm long, base dorsally gibbous, above gibbosity 0.3-0.5 cm wide, middle slightly ventricose, throat slightly contracted, 0.5-0.8 cm wide, outside villous, inside glabrous, limb 1-2 cm wide, lobes unequal, upper lobes connate into a galea, erect, ca. 0.7 x 0.5-0.6 cm, margin entire, lateral lobes erect, rounded to truncate, 0.1-0.2 x 0.3-0.5 cm, margin entire, basal lobe erect, broadly rounded, 0.2-0.3 x 0.3-0.6 cm, margin entire; stamens slightly exserted, inserted at base of corolla tube; ovary ovoid, 0.4-0.5 x 0.3-0.4 cm, pubescent, style 4-5 cm long, puberulous, stigma stomatomorphic. Mature capsule brownish, ovoid, 1-1.2 x 0.5-0.7 cm.

Distribution: Northern Brazil (Amazonas, Roraima) and Guyana; growing on granitic outcrops or in wet depressions on sandstone, at 120-1400 m alt.; 15 specimens examined (GU: 13).

Specimens examined: Guyana: Potaro-Siparuni Region, summit of Malakwalai-Tipu, Henkel *et al.* 5546 (BRG, NY, U, US); Upper Takatu-Upper Essequibo Region, Kanuku Mts., Iraimakipang summit, Goodland & Maycock 456B (US), Goodland & Persaud 704 (US), A.C. Smith 3652 (F, GH, K, NY, P, U, US); Kanuku Mts., Grewal *et al.* 342 (U); Rupununi, Harrison 1346 (K, NY).

Phenology: Collected in flower April-September, in fruit in July and September.

Note: Hanstein (1865, p. 270) cited a Poiteau collection from French Guiana belonging to this species, but none has been located that can be identified as this species. There is, however, a Poiteau specimen at K which can be identified as *Sinningia incarnata*.

20. **TYLOPSACAS** Leeuwenb., Taxon 9: 220. 1960. – *Tylosperma* Leeuwenb., Acta Bot. Neerl. 7: 323. 1958, non Botsch. 1952, nor Donk, 1957.

Type: T. cuneata (Gleason) Leeuwenb. (Episcia cuneata Gleason)

Terrestrial acaulescent erect herbs, without modified stems. Stems unbranched. Leaves radicular, rosulate, equal to strongly unequal in a pair, venation pinnate, foliar nectaries absent. Flowers axillary, few to numerous in pedunculate panicles; bracteoles absent; pedicellate. Calyx lobes connate only at base; corolla white, tubular; stamens included, filaments not connate, anthers coherent in pairs, dehiscing by longitudinal slits, thecae divergent at base; staminode small; disc a ring of 5 subequal glands or with 2 dorsal lobes much larger than the other 3; ovary superior, stigma capitate. Fruit a dry, brown, loculicidally and sometimes septicidally dehiscent, 2-4-valved capsule, valves opening slightly.
Chromosome number unknown.

Distribution: A monospecific genus, known only from the Guayana Highlands in Bolívar and Amazonas in Venezuela, neighbouring Brazil, western Guyana, and neighbouring Brazil; growing among mossy rocks, near waterfalls in mixed forest.

1. **Tylopsacas cuneata** (Gleason) Leeuwenb., Taxon 9: 221. 1960. – *Episcia cuneata* Gleason, Bull. Torrey Bot. Club 58: 467. 1931. – *Tylosperma cuneatum* (Gleason) Leeuwenb., Acta Bot. Neerl. 7: 323, 432, fig. 33. 1958. Type: Venezuela, Amazonas, Tate 879 (holotype NY, isotypes K, US). – Fig. 15 E-H

Terrestrial herb, 10-30 cm tall (including leaves). Stem succulent, erect, villous. Leaves equal to strongly unequal in a pair; petiole to 9 cm long, appressed tomentose to hirsute; blade membranous or papyraceous when dry, oblanceolate, larger blade in a pair 2-28 x 0.8-5 cm, margin sharply serrate, apex acute to acuminate, base cuneate to long-decurrent, above hirsute or subtomentose to glabrous, below subtomentose to glabrous. Flowers numerous, paniculate; peduncle 1-10 cm long, sparsely strigillose; pedicel 0.5-3.5 cm long, sparsely strigillose. Calyx subcampanulate, green, lobes shortly connate at base, erect to spreading, subequal, linear-lanceolate, 0.3-0.6 x 0.05-0.1 cm, margin entire, apex long acuminate, outside sparsely strigillose, inside glabrous; corolla erect in calyx, white, 0.8-1 cm long, tube cylindric, 0.65-0.75 cm long, slightly constricted above base, 0.1-0.15 cm, middle funnelform, throat not ventricose, 0.15-0.2 cm wide, outside glabrous, inside glabrous, limb 0.5-0.7 cm wide, lobes subequal, spreading, ovate-oblong, 0.2-0.3 x 0.15-0.2 cm, margin entire, glandular-pubescent; stamens included, adnate for ca. 0.2 cm to base of corolla tube; ovary ovoid, 0.2 x 0.15 cm, puberulent to glabrous, style 0.25-0.3 cm long, glabrous, stigma capitate. Mature capsule green to brown, oblong-ovoid, 0.3-0.5 x 0.2 cm.

Distribution: Eastern Venezuela (Amazonas, Bolívar), northern Brazil (Amazonas, Pará) and Guyana; growing in low and high mixed forests, among mosses on steep moist rocks near rivers and waterfalls, at 80-1250 m alt.; 65 collections studied (GU: 13).

Specimens examined: Guyana: Cuyuni-Mazaruni Region, Makreba Falls, Kurupung R., Altson 357 (BRG, NY), Pinkus 11 (GH, NY, US); Pakaraima Mts., Imbaimadai Savannas, Maipuri Falls, Karaurieng R., Maguire & Fanshawe 32280 (NY, US); Potaro-Siparuni Region, Potaro R. Gorge, Johnson's View down to base of Kaieteur Falls, Cowan & Soderstrom 1925 (NY, SEL, US); Kaieteur Falls, Kvist *et al.* 363 (AAU, B, BBS, BRG, CAY, COL, NY, P, U, US); Potaro R. Gorge, Maguire & Fanshawe 23520 (A, BR, F, G, K, MO, NY, P, U, UC, US, VEN); Mt. Kanaima, Whitton 154 (K).

Phenology: Collected in flower in February, April, May, July, August, and October, in fruit in October.

TAXONOMIC AND NOMENCLATURAL CHANGES, NEW TYPIFICATIONS

New synonyms:
Besleria maasii Wiehler to Besleria patrisii DC.
Besleria verecunda C.V. Morton to Besleria patrisii DC.
Chrysothemis villosa (Benth.) Leeuwenb. to Chrysothemis pulchella (Donn ex Sims) Decne.
Columnea aureonitens Hook. to Columnea sanguinea (Pers.) Hanst.
Columnea calotricha var. *austroamericana* C.V. Morton to Columnea calotricha C.V. Morton
Columnea calotricha var. *breviflora* C.V. Morton to Columnea calotricha C.V. Morton
Columnea steyermarkii C.V. Morton to Alloplectus savannarum C.V. Morton

Lectotypifications:
Besleria flavovirens Nees & Mart.
Besleria penduliflora Fritsch
Rechsteineria crenata Fritsch

Neotypification:
Napeanthus jelskii Fritsch

NUMERICAL LIST OF ACCEPTED TAXA

1. Alloplectus Mart.
   1-1. A. cristatus (L.) Mart. var. epirotes Leeuwenb.
   1-2. A. savannarum C.V. Morton

2. Besleria L.
   2-1. B. flavovirens Nees & Mart.
   2-2. B. insolita C.V. Morton
   2-3. B. laxiflora Benth.
   2-4. B. parviflora L.E. Skog & Steyerm.
   2-5. B. patrisii DC.
   2-6. B. penduliflora Fritsch
   2-7. B. saxicola C.V. Morton

3. Chrysothemis Decne.
   3-1. C. pulchella (Donn ex Sims) Decne.
   3-2. C. rupestris (Benth.) Leeuwenb.

4. Codonanthe (Mart.) Hanst.
   4-1. C. calcarata (Miq.) Hanst.
   4-2. C. crassifolia (Focke) C.V. Morton

5. Codonanthopsis Mansf.
   5-1. C. dissimulata (H.E. Moore) Wiehler

6. Columnea L.
   6-1. C. calotricha Donn. Sm.
   6-2. C. guianensis C.V. Morton
   6-3. C. oerstediana Klotzsch ex Oerst.
   6-4. C. sanguinea (Pers.) Hanst.
   6-5. C. scandens L.

7. Corytoplectus Oerst.
   7-1. C. deltoideus (C.V. Morton) Wiehler

8. Cremersia Feuillet & L.E. Skog
   8-1. C. platula Feuillet & L.E. Skog

9. Drymonia Mart.
   9-1. D. antherocycla Leeuwenb.
   9-2. D. coccinea (Aubl.) Wiehler
   9-3. D. psilocalyx Leeuwenb.
   9-4. D. serrulata (Jacq.) Mart.

10. Episcia Mart.
    10-1.   E. reptans Mart.
    10-2.   E. sphalera Leeuwenb.
    10-3.   E. xantha Leeuwenb.

11. Gloxinia L'Hér.
    11-1.   G. perennis (L.) Fritsch
    11-2.   G. purpurascens (Rusby) Wiehler

12. Kohleria Regel
    12-1.   K. hirsuta (Kunth) Regel var. hirsuta

13. Lampadaria Feuillet & L.E. Skog
    13-1.   L. rupestris Feuillet & L.E. Skog

14. Lembocarpus Leeuwenb.
    14-1.   L. amoenus Leeuwenb.

15. Napeanthus Gardner
    15-1.   N. angustifolius Feuillet & L.E. Skog
    15-2.   N. jelskii Fritsch
    15-3.   N. macrostoma Leeuwenb.
    15-4.   N. rupicola Feuillet & L.E. Skog

16. Nautilocalyx Linden ex Hanst.
    16-1.   N. adenosiphon (Leeuwenb.) Wiehler
    16-2.   N. bryogeton (Leeuwenb.) Wiehler
    16-3.   N. coccineus Feuillet & L.E. Skog
    16-4.   N. cordatus (Gleason) L.E. Skog
    16-5.   N. fasciculatus L.E. Skog & Steyerm.
    16-6.   N. kohlerioides (Leeuwenb.) Wiehler
    16-7.   N. mimuloides (Benth.) C.V. Morton
    16-8.   N. pallidus (Sprague) Sprague
    16-9.   N. pictus (Hook.) Sprague
    16-10.   N. porphyrotrichus (Leeuwenb.) Wiehler
    16-11.   N. punctatus Wiehler

17. Paradrymonia Hanst.
    17-1.   P. anisophylla Feuillet & L.E. Skog
    17-2.   P. barbata Feuillet & L.E. Skog
    17-3.   P. campostyla (Leeuwenb.) Wiehler
    17-4   P. ciliosa (Mart.) Wiehler
    17-5.   P. densa (C.H. Wright) Wiehler
    17-6.   P. longifolia (Poepp.) Wiehler
    17-7.   P. maculata (Hook. f.) Wiehler

18. Rhoogeton Leeuwenb.
    18-1.   R. cyclophyllus Leeuwenb.
    18-2.   R. viviparus Leeuwenb.

19. Sinningia Nees
    19-1.   S. incarnata (Aubl.) D.L. Denham
    19-2.   S. schomburgkiana (Kunth & Bouché) Chautems

20. Tylopsacas Leeuwenb.
    20-1.   T. cuneata (Gleason) Leeuwenb.

COLLECTIONS STUDIED
(Numbers in **bold** represent types)

GUYANA

Abraham, A.A., 340 (16-9); 345 (2-7)

Altson, R.A., 311 (2-7); 321 (16-4); 357 (20-1); 371 (16-2); 422 (16-9)

Andel, T. van, *et al.*, 1075 (16-7); 1739 (4-2); 1954 (4-2)

Anonymous, **s.n.** (17-7)

Appun, C.F., s.n. (16-9); 66 (3-2); 2125 (19-1); 5181 (19-1)

Archer, W.A., 2321 (16-7); 2432 (17-7)

Atkinson, D.J., 33 (16-4); 77 (17-5)

Bailey, I.W., 110 (4-1); 162 (16-9); **181** (4-1)

Bartlett, A.W., 8562 (9-4); 8743 (2-2)

Beckett, J.E., s.n. (17-7)

Beddington, H., 33 (2-3)

Boom, B.M., *et al.*, 7484 (1-2); 7734 (18-2); 7746 (1-2); 8838 (16-9); 8843 (11-2); 8909 (1-2); 8910 (6-2); 8990 (17-1); 9150 (2-6); 9156 (18-1); **9186** (16-3); 9187 (18-1); 9197 (1-1); **9202** (15-4)

Boyan, R., RB135 (3-1)

BW (Boschwezen), 6897 [Persaud 140] (17-7)

Chanderbali, A. & D. Gopaul, 81 (17-5)

Christenson, E.A., *et al.*, 1917 (17-5)

Clarke, H.D. [D.], *et al.*, 250 (4-2); 319 (4-1); 443 (4-2); 444, 599 (4-1); 615 (2-3); 638 (17-5); 645 (4-1); 725 (17-5); 750 (4-1); 856 (16-7); 1072 (1-2); 1138 (12-1); 1322 (4-2); 1439 (16-4); 1495, 1515 (4-1); 1790 (19-2); 1820 (3-2); 1928 (17-7); 1929 (9-2); 1959 (4-2); 1973, 2008 (4-1); 2196 (9-2); 2440 (3-1); 2452 (19-1); 2520 (17-5); 2525 (2-3); 2596, 2653 (4-1); 2844 (9-2); 2876 (16-9); 2931, 2932 (2-3); 2961 (16-9); 3056, 3454 (4-2); 3646 (9-2); 3667, 3907 (4-1); 4112 (16-4); 4181 (17-7); 4213 (4-1); 4214 (10-3); 4258 (2-3); 4260 (10-3); 4369 (9-2); 4392 (16-11); 4408 (4-2); 4454, 4616 (9-2); 4619, 4693 (3-2); 4770 (4-2); 4908 (3-2); 5157 (19-1); 5312 (1-2); 5434 (6-2); 5523 (2-6); 5526 (7-1); 5549 (2-6); 5616 (1-2); 6007 (16-7); 6475 (9-2); 7494 (2-3); 7705 (16-9); 7818 (9-2); 7998 (16-11); 8102, 8158 (9-2); 8185 (12-1); 8431 (9-2); 8564 (16-5); 8897 (13-1); 8950, 9006, 9202 (17-1); 9255 (18-2); 9256, 9444 (16-3); 9618, 9619 (17-1)

Cowan, R.S., 39228 (17-5); 39337 (17-7); 39358 (4-1)

Cowan, R.S. & T.R. Soderstrom, 1751 (1-2); 1837 (5-1); 1851 (16-4); 1925 (20-1); 2123 (4-1); 2148 (18-2); 2149 (18-1); 2208 (17-4)

Cruz, J.S. de la, 1135, 1316 (4-1); 1405 (4-2); 1518, 1535 (17-7); 1584 (16-9); 1588, 1787 (10-1); 2139 (17-5); 2197 (4-2); 2294 (4-1); 2327 (16-9); 2350 (10-1); 2810 (4-2); 3022, 3124 (17-7); 3250

Im Thurn, E., s.n. (19-1); 287 (2-6)

Irwin, H.S., BG-62 (4-1)

Jansen-Jacobs, M.J., *et al.*, 695 (4-1); 1225 (3-2); 1670 (4-2); 1764 (4-1); 2316 (9-2); 2417 (4-1); 2476, 2879 (4-2); 3497 (3-2); 3853 (4-2); 3964, 4104 (19-1); 4319 (3-2); 4429 (3-1); 4432 (19-2); 4848 (19-1); 5629 (3-2); 5821 (16-11); 5875 (9-2); 5899 (3-2); 5987 (4-1)

Jenman, G.S., **880** (18-2); 896 (18-1); 2364 (16-9); **2414** (17-5); 3990 (3-1); **4156** (2-5); 5457 (16-9); 5608 (3-1); 7123 (16-7)

Kelloff, C.L., *et al.*, 867 (1-2); 1040 (16-4); 1402 (17-4)

Knapp, S. & J. Mallet, 2908 (4-1)

Kvist, L.P., *et al.*, 34 (1-2); 35 (16-4); 197 (16-9); 214 (6-2); 217 (4-1); 245 (11-2); 256 (1-2); 363 (20-1); 364 (6-2); 368 (16-4); 370 (18-2)

Lance, K., *et al.*, 3 (16-9); 8 (16-4); 53 (1-2)

Lang, H., 113, 155 (16-9)

Linder, D.H., 7 (16-9)

Maas, P.J.M., *et al.*, 2558 (1-2); 2621 (16-4); 3555, 3595 (4-2); 3930 (2-3); 3949, 4125 (4-1); 4336, 4441 (16-4); 5447 (4-2); 5599 (16-7); 5628 (16-10); 5634 (4-1); 5869 (17-5); 5903 (16-9)

Maguire, B., *et al.*, 22826 (17-7); 22838 (4-1); 22982 (17-5); 23019 (17-4); 23036 (17-5); 23047 (16-9); **23067** (6-2); 23078 (16-4); **23127** (1-2); 23520 (20-1); 32088 (17-5); 32280 (20-1); 32343 (1-2); 32366 (16-4); 32368 (17-4); 32387 (16-4); 32392 (16-2);

40488 (16-7); **40585** (18-1); 40586a (16-3); 40586b, 40588 (16-4); **40594** (1-1); 40595 (1-2); 45984A, 46051A (18-2)

McDowell, T.D., *et al.*, 2246 (4-2); 2596 (16-7); 3051 (3-1); 3054 (16-10); 3415 (4-2); 3423 (2-7); 3453 (16-9); 3638 (4-1); 3663 (4-2); 3664 (4-1); **3810** (17-2); 3823 (20-1); 3851 (2-7); 3860 (16-9); 3917 (17-4); 3919 (16-9); 3951, 3966 (16-4); 3984 (6-2); 4025 (1-2); 4182 (17-7); 4237 (4-2); 4240 (16-7); 4264 (4-1); 4379 (2-7); 4418 (16-7); 4476 (2-1); 4721 (1-2); 4808 (4-2); 4819 (17-7); 4820 (17-5); 4823 (16-9); **4872** (13-1); 4910 (1-2); 4915 (2-7); 4923 (20-1)

Mell, C.D. & R.C. Mell, 200 (16-9)

Mori, S.A., *et al.*, 8104 (16-9); 8184 (17-5); 24647 (10-3)

Mutchnick, P., *et al.*, 136 (2-1); 214 (4-1); 301 (2-4); 349 (12-1); 405, 544, 792, 1126 (4-1); 1129, 1231, 1286, 1328 (4-2); 1623 (12-1); 1625 (19-2)

Myers, J.G., 5839 (2-3)

Parker, s.n. (19-1)

Persaud, A.C., 350 (16-9)

Persaud, C.A., 140 [BW 6897] (17-7)

Persaud, R., 26 (1-2); 79 (18-2); 80 (18-2); 84 (7-1); 131 (6-2)

Peterson, P. & D. Gopaul, 7654 (19-1)

Pinkus, A.S., 11 (20-1); **12** (16-2); 244 (4-1)

Pipoly, J.J., *et al.*, 7320 (4-2); 7522, 7572 (17-5); 8059 (17-7); 8298 (16-7); 8330 (16-7); 8818 (4-2); 9683 (4-1); 9951

Donselaar, J. van, 1463, 2115 (4-2); 2541 (9-2); 2938 (4-1)

Donselaar, J. van & J.P. Schulz, LBB 10566 (3-2)

Elburg, J.P., LBB 13501 (9-2)

Evans, R.J., *et al.*, 2437, 2877 (9-2); 2972 (9-1); 3027 (16-9)

Florschütz, J. & P.A. Florschütz, 359 (9-2); 515 (4-2); 613, 1181 (4-1); 1326, 1400 (3-2); 1522 (4-2); 2213 (3-1)

Florschütz, P.A. & P.J.M. Maas, 2657 (4-1); 2798 (3-2); 2830, 2858 (9-2); 2890 (2-5); 3078 (17-3)

Focke, H.C., **s.n.** (4-2); **766** (9-4); **822** (11-1); **941** (4-1); **975** (9-2)

Gieteling, C.J., 42 (9-4)

Gonggrijp, J.W., 14 [BW 4115] (9-2); 171 [BW 2180] (9-4)

Gonggrijp, J.W. & G. Stahel, 182 (2-3); 183 (15-3)

Granville, J.J. de, *et al.*, 958 (9-3); 1104 (16-5); 1482 (15-3); 1484 (9-2); 12067 (17-4); 12100 (16-5); 12107 (16-7); 12147 (9-3); 12165 (3-1); 12261 (15-3)

Hammel, B.E., *et al.*, 21373 (17-3); 21461, 21583 (9-1); 21742 (9-2)

Hawkins, T., 1888 (14-1)

Holmgren, N., *et al.*, 54396 (1-2)

Hostmann, F.W.R., 2044, 2045 (16-4)

Hostmann, F.W.R. & A. Kappler, 1372 (9-2)

Hostmann, F.W.R., *et al.*, B949 (9-4)

Hulk, J., 243 (3-2)

Indigenous collector, 99 (9-2); 255 (4-1); 278 (9-4); 279 (17-3)

Irwin, H.S., *et al.*, 54558 (3-2); 54641 (2-5); 54709 (2-5); 54766 (6-4); 54873 (1-2); 54917 (16-9); 54959 (1-2); 54963 (16-9); 55100 (6-4)

Jonker, F.P. & A.M.E. Jonker, 308, 314 (4-2); 486 (9-2); **625** (17-3)

Kappler, A., **2044** (10-2)

Kock, C., s.n. (17-3)

Koster, J.T., LBB 13025 (17-3); LBB 13026 (9-2)

Kramer, K.U., *et al.*, 2118 (4-2); 2449 (9-2); 2451 (4-2); 2733, 3046 (4-1); 3076, 3251 (14-1)

Kuyper, J., 1 (4-2)

Lanjouw, J., 1244 (4-1)

Lanjouw, J. & J.C. Lindeman, 1410, 1940 (4-2); 2096 (9-2); 2104 (4-2); 2373 (2-1); 2466, 2593 (14-1); 2630 (15-3); 2686 (16-9); 2742 (6-1); 2833 (14-1); 2890, 2941 (16-9)

LBB (Lands Bosbeheer), 8453 (9-2); 8712, 8712 (9-4); 8717 (4-2); 9104 (9-4); 9160, 9578 (9-2); 10209, 10256a (2-5); 10313 (14-1); 10452 (2-5); 10566, 10566 (3-2); 10681 (9-4); 10756 (4-2); 10876 (9-2); 10909 (17-3); 10961 (3-2); 12075 (2-3); 12133 (9-2); 12558 (6-1); 12782 (9-4); 13025 (17-3); 13026 (9-2); 13452 (3-2); 13501, 14664 (9-2); 14985 (9-4); 16239 (4-1)

Lindeman, J.C., *et al.*, 350 (4-2); 3613 (4-2); 4987 (9-2); 5186 (4-2); 6814 (9-4); LBB 12075 (2-3); LBB 12133 (9-2)

Lindeman, J.C. & A.R.A. Görts-van Rijn *et al.*, 151 (4-2); 323 (3-1)

Lindeman, J.C. & E.A. Mennega, 24 (2-3); 42 (9-2); 152 (16-1)

Zaandam, C.J., s.n. [BW 6323], s.n. [BW 6328] (9-2); s.n. [BW 6619] (6-1)

FRENCH GUIANA

Acevedo, P., *et al.*, 4806, 4908, 4969 (9-2); 4997 (15-3)
Allorge, L., 378 (10-3); 392 (9-2)
Andersson, L., *et al.*, 1925 (6-3)
Anonymous, s.n. (15-1); s.n. (2-5)
Aublet, J., **s.n.** (9-2); s.n. (2-1); s.n. (16-9); **s.n.** (19-1); s.n. (6-3); s.n. (2-5)
Aubréville, A., 65, 264 (9-2); 348 (6-3); 368 (9-2)
BAFOG, 4472 (9-2); 4536 (2-5)
Barrier, S., *et al.*, 2519 (15-3); 2530 (6-4); 2543 (9-2); 2610 (9-4); 2624 (6-3); 2625 (4-2); 2637 (6-3); 2660 (9-2); 2692, 2700, 2738 (6-3); 2746 (9-2)
Belbenoit, P., 193N (9-2)
Benoist, R., s.n. (11-1); 57 (4-1); 194 (4-2); 196 (9-2); 820 (16-9); 1240 (2-5)
Billiet, F., *et al.*, 684 (2-5); 1058 (4-2); 1109 (4-1); 1122 (9-2); 1209 (4-2); 1315 (16-9); 1583 (4-1); 1600 (16-9); 1684 (2-5); 1685 (9-2); 1823 (16-9); 1856 (17-5); 1857 (17-7); 1918 (15-2); 1920 (2-2); 2006 (17-5); 2007 (4-1); 2033 (9-3); 2034 (16-7); 4421, 4567 (9-2); 5750 (17-5); 5800 (6-4); 5818 (15-3); 6256 (2-2); 6269 (6-3); 6271 (15-2); 6281 (16-9); 6358 (9-2); 6378 (15-1); 6384 (17-5); 6435 (2-5); 6438 (15-3); 6439 (2-1); 6465 (6-3); 7113 (4-1); 7406 (9-2); 7463 (15-2); 7468 (10-3)

Bitaillon, C., 43, 50 (16-9); 138 (9-2)
Blanc, M., 162, 178 (9-2); 207 (17-3)
Blanc, P., *et al.*, 85-124 (16-1); 93-4 (10-2); 93-100 (16-7); 93-102 (2-2)
Boggan, J.K., *et al.*, 110 (9-4)
Boom, B.M., *et al.*, 1564, 10751 (10-3); 10791 (6-3); 10818 (9-2)
Bordenave, B., 298, 471 (9-2); 634, 727 (16-9); 1236 (4-1); 2404 (17-5); 2525, 2778, 2862 (9-2); 2880 (16-9)
Chapuis, J., 30 (9-2)
Christenson, E.A. & S.R. George, 1859 (16-9)
Chuah, M., 138 (9-4); 162 (9-2); 166 (6-4)
Cosson, H.E., 18 (9-2)
Cowan, R.S., *et al.*, 38725 (17-5); 38734 (9-2); 38756 (6-3); 38776 (2-2); 39219 (16-9)
Cremers, G., *et al.*, 3965 (9-2); 4232, 4288 (4-2); 4514 (11-2); 4617, 4641 (6-1); 5017 (9-2); 5101 (4-2); 5105, 5122 (4-1); 5397 (6-3); 6090 (10-3); 6336 (16-9); 6380 (15-3); 6420 (6-3); 6527 (2-5); 6925 (5-1); 6969 (9-2); 7064 (10-3); 7164 (17-5); 7567 (16-1); 8162 (2-5); 8200 (9-2); 8599 (2-2); 8600 (16-9); 8630 (2-1); 9676 (9-2); 9903 (6-1); 9951 (17-4); 10884a (15-2); 10884b (15-3); 10891 (2-5); 10898 (9-2); 11096 (4-2); 11319 (9-2); 11342 (17-3); 11391 (16-9); 11476 (9-2); 11492 (4-2); 11771 (17-5); 11951, 11989 (16-9); 12030 (17-4); 12057 (17-5); 12070

7516 (6-4); 7535 (15-2); 7549 (16-9); 7603 (9-3); 7619 (6-3); 7657 (16-9); 7688 (2-5); 7719 (15-3); 7812 (2-1); 7851 (10-3); 7881 (15-2); 7939 (6-1); 7958 (16-9); 7977 (9-2); 8073 (6-3); 8143 (2-5); 8271 (10-2); 8429 (9-2); 8461 (2-5); 8482 (15-3); 8517 (2-1); 8549 (6-3); 8585 (9-3); 8628 (6-4); 8657 (16-9); 8663 (2-5); 8733 (17-3); 8833 (10-3); 8843 (2-5); 8847 (4-2); 9021 (9-3); 9090 (2-2); 9444 (3-1); 9768 (9-2); 9780 (19-1); 9919, 9962 (19-1); 9974 (16-9); 10080 (4-2); 10347 (17-5); 10383, 10464 (9-2); 10506 (2-5); 10553 (4-2); 10573 (9-3); 10607 (4-2); 10626 (2-2); 10637 (15-3); 10638 (10-3); 10644 (16-9); 10665 (9-2); 10675 (6-3); 10729 (15-3); 10730 (2-1); 10834 (9-3); 10871 (6-3); 10895 (6-1); 10949 (2-5); 10969 (9-4); 11033 (14-1); 11063 (16-7); 11112 (2-5); 11123 (15-3); 11132 (16-9); 11177, 11178 (14-1); 11192 (17-5); 11259, 11260 (16-9); 11270 (17-3); 11351 (2-2); 11562 (9-2); 11739 (19-1); 11752 (17-3); 11791 (6-3); 11837 (17-3); 12692 (2-1); 12713 (15-2); 12727 (15-3); 12729 (2-5); 12730 (2-1); 12805 (16-9); 12841 (15-3); 12841 (15-2); 12908 (9-4); 12971 (10-2); 13009 (17-7); 13031 (16-6); 13038 (17-3); 13153 (4-1); 13205 (10-3); 13223 (2-2); 13271 (6-3); 13327 (17-3); 13346 (15-3); 13352 (2-1); 13355 (2-5); 13360 (9-2); 13361 (9-3); 13370 (16-9); 13514 (9-2); 13533 (2-5); 13538 (9-4); 13555 (10-2); 13845 (17-3); 13879 (2-5); 13898 (9-3); 13908 (2-5); 13921 (15-3); 13942 (6-1); 13954 (16-9); 13959 (16-1); 13994 (9-2); 14088 (16-9); 14226 (17-3); 14431 (9-3); 14436 (6-3); 14532 (2-2); 14545 (16-9); 14682 (4-2); 14868 (8-1)

Grenand, P., et al., 51 (9-2); 98 (9-4); 309 (16-6); 310 (6-1); 355 (16-6); 713 (9-4); 896 (17-3); 1996 (16-8); 2003 (17-4); 1215 (14-1); 1876 (9-4); 2851 (16-9)

Hahn, W.J., 3617 (2-5); 3738 (9-2)

Hallé, F., 59 (9-2); 79 (16-9); 560 (9-2); 562 (4-1); 696 (6-1); 776 (10-3); 787 (9-2); 1131, 2302 (6-3); 2304 (6-4); 2314 (2-5); 2315 (2-1); 2316, 2457 (15-3); 2460, 2668 (2-2); 2902 (16-9)

Haxaire, C., 290 (9-2); 692, 834, 877 (16-6)

Hequet, V., 172, 183 (9-2); 214 (9-4); 420 (15-2); 453 (9-2); 514, 531 (9-4); 550 (4-1); 573 (4-2)

Herb. Exposition Coloniale, s.n. (15-3)

Herb. Maire, s.n. (15-1); s.n. (9-2); s.n. (16-7); s.n. (2-1)

Herb. Richard, s.n. (2-5)

Herb. Rudge, s.n. (9-4)

Hoff, M., et al., 5924 (9-2); 6284, 6303 (2-2); 6315 (17-3); 6316 (17-5); 6367 (16-9); 6542 (17-5); 6555 (16-9);

4); T-356 (9-2); B-439 (16-9); T-507 (9-4); B-534 (16-9); T-549 (9-2); B-552 (6-1); T-568 (16-7); T-571 (17-3); T-576 (6-4); T-637 (6-1); T-667 (16-6); B-689 (9-2); T-690 (17-3); B-693 (4-2); B-694 (16-9); T-694 (9-2); T-708 (6-1); T-709 (6-4); T-796 (9-2); T-800 (17-3); T-808 (9-2); B-826 (2-5); B-842 (9-2); B-896 (9-4); T-943 (9-2); T-948 (16-6); B-1061 (6-3); 1157 (4-2); 1177 (6-1); 1178, 1244 (6-3); 1322 (4-2); 1419 (6-3); B-1521 (6-1); B-1554 (9-4); 1573 (9-2); B-1672 (16-9); 1685 (4-1); 1769 (4-2); B-1775 (17-3); 1778 (4-2); 1805 (9-2); B-1809 (17-3); B-1816 (2-5); B-1818, B-1837, B-1858, 1860 (9-2); B-1908 (4-1); B-1984 (19-1); B-1988 (9-2); 2024 (2-2); B-2079 (6-3); B-2109 (4-1); B-2155 (9-2); B-2158 (6-3); B-2159 (2-2); B-2160 (2-5); B-2174 (6-1); B-2205 (9-4); B-2288 (16-9); 2293 (9-2); 2336 (4-2); 2354 (6-3); B-2355 (2-2); B-2368 (17-5); 2371 (9-4); B-2376 (10-3); B-2391 (9-4); B-2578 (10-2); 2612 (9-2); 2687, 2695 (4-1); 2722 (9-2); 2743 (6-1); B-2746 (9-4); 2757 (10-2); 2814, 2916 (9-2); 2935 (9-4); 2959 (6-1); 2978 (2-2); 3001 (6-1); 3027 (4-1); B-3033 (9-2); 3057 (9-4); 3061 (6-1); B-3109 (16-6); 3127 (6-4); B-3147 (9-4); B-3160, B-3160 (4-1); B-3169 (16-6); B-3217 (6-1); B-3242 (4-2); B-3287 (9-4); B-3300, B-4036 (9-2); B-4351 (2-5)

Pasch, 9161 (9-2)

Patris, **s.n.** (2-5); **s.n.** (9-2)

Perrottet, G.S., s.n. (19-1); s.n. (2-5)

Petitbon, J., 157 (9-4)

Philippe, M., 174, 293 (16-9); 319 (15-2)

Phillippe, L.R., et al., 26916 (9-2); 27024 (15-3)

Pignal, M. & O. Poncy, 730 (9-2)

Pipoly, J.J., et al., 11017 (16-9)

Plaige, V., 10 (9-2)

Poiteau, P., s.n. (19-1); s.n. (9-2); s.n. (2-2); s.n. (6-4); s.n. [in 1826], s.n. [in herb. J.E. Gay] (15-1); s.n. (11-1)

Poncy, O., et al., 127 (4-2); 159, 160 (2-5); 847 (15-2); 855 (10-3); 856 (16-7); 1056 (10-3); 1205 (17-3); 1416 (9-3); 1422 (6-3)

Prance, G.T., et al., 30638 (2-2)

Prévost, M.F., et al., 208 (2-2); 322 (9-2); 471, 532 (4-2); 545, 569 (16-9); 578 (17-5); 896 (17-3); 1239 (2-2); 1268 (2-1); 1409 (9-2); 1517 (16-9); 1601 (16-7); 1688 (4-2); 1740 (17-5); 1767 (2-2); 1793 (2-1); 1827 (2-5); 1828 (15-3); 1833 (9-4); 1863 (16-9); 1885 (6-1); 1892 (9-4); 1903 (17-4); 1909 (16-8); 1912 (16-7); 1988 (17-3); 1996 (17-7); 2000 (17-3); 2013 (6-1); 2066 (9-4); 2223 (14-1); 2253 (16-7); 2337 (6-1); 3509 (10-3); 3770 (9-2); 3887 (10-3); 4161 (16-9)

Profizi, J.P., H35 (9-2)

Puig, H., 12026 (2-5)

Raynal-Roques, A., 19976 (9-2)

Reynolds, J. & L. Brothers, 163 (16-7)

INDEX TO SYNONYMS, NAMES IN NOTES AND SOME TYPES

Chrysothemis
  aurantiaca Decne. = 3-1
  villosa (Benth.) Leeuwenb. = 3-1
Clerodendrum
  verrucosum Splitg. ex de Vriese, see 2, note
Codonanthe
  sect. *Codonanthopsis* (Mansf.) H.E. Moore = 5
  bipartita L.B. Sm. = 4-1
  confusa Sandwith = 4-2; see 4-2, note
  dissimulata H.E. Moore = 5-1
  gracilis (Mart.) Hanst., see 4, type
Codonanthopsis
  ulei Mansf., see 5, type
Collandra
  aureonitens (Hook.) Hanst. = 6-4
  picta (Hook.) Lem. = 16-9
  sanguinea (Pers.) Griseb. = 6-4
Columnea
  acuminata Benth., see 6, type
  anisophylla DC., see 6, type
  aurantiaca Decne. & Planch., see 6, type
  aureonitens Hook. = 6-4; see 6-4, note
  calotricha Donn. Sm., see 16-6, use
  calotricha Donn. Sm. var. *austroamericana* C.V. Morton = 6-1
  calotricha Donn. Sm. var. *breviflora* C.V. Morton = 6-1
  ciliosa (Mart.) Kuntze = 17-4
  coccinea (Aubl.) Kuntze = 9-2
  cristata (L.) Kuntze = 1-1
  minor (Hook.) Hanst., see 6, type
  patrisii (DC.) Kuntze = 9-2
  picta (Hook.) Hanst. = 16-9
  sanguinea (Pers.) Hanst., see 6, type
  steyermarkii C.V. Morton = 1-2
  strigosa Benth., see 6, type
Corytoplectus
  capitatus (Hook.) Wiehler, see 7, type
Crantzia Scop. = 1, see 1-1, note
  coccinea (Aubl.) Fritsch = 9-2
  cristata (L.) Fritsch = 1-1; see 1-1, note; see 1, type
  epirotes (Leeuwenb.) J.L. Clark, see 1-1, note
  patrisii (DC.) Fritsch = 9-2
Cyrtodeira Hanst. = 10
  cupreata (Hook.) Hanst., see 10, type

Dalbergaria Tussac = 6
    aureonitens (Hook.) Wiehler = 6-4
    guianensis (C.V. Morton) Wiehler = 6-2
    phaenicea Tussac, see 6, type
    sanguinea (Pers.) Steud. = 6-4
Drymonia
    calcarata Mart. = 9-4; see 9, type
    campostyla Leeuwenb. = 17-3
    cristata Miq. = 9-4
    longifolia Poepp. = 17-6
    psila Leeuwenb. = 9-3
    serrulata (Jacq.) Mart., see 9, type
Episcia
    sect. *Centrosolenia* (Benth.) Benth. & Hook. f. = 16
    sect. *Nautilocalyx* (Linden ex Hanst.) Benth. & Hook. f. = 16
    sect. *Pagothyra* Leeuwenb. = 17
    sect. *Paradrymonia* (Hanst.) Leeuwenb. = 17
    sect. *Salpinganthus* Leeuwenb. = 17
    sect. *Trichosperma* Leeuwenb. = 16
    subsect. *Centrosolenia* (Benth.) Leeuwenb. = 16
    subsect. *Tremanthera* Leeuwenb. = 10
    adenosiphon Leeuwenb. = 16-1
    bryogeton Leeuwenb. = 16-2; see 16, type
    ciliosa (Mart.) Hanst. = 17-4
    cordata Gleason = 16-4
    cuneata Gleason = 20-1; see 20, type
    cupreata (Hook.) Hanst., see 10, note
    densa C.H. Wright = 17-5; see 17, type
    glabra (Benth.) Hanst. = 17-4
    hirsuta (Benth.) Hanst. = 16-4
    kohlerioides Leeuwenb. = 16-6
    lilacina Hanst., see 10, note
    longifolia (Poepp.) Hanst. = 17-6
    maculata Hook. f. = 17-7; see 17, type
    melittifolia (L.) Mart., see 16, note
    mimuloides Benth. = 16-7
    picta (Hook.) Hanst. = 16-9
    porphyrotricha Leeuwenb. = 16-10
    pulchella (Donn ex Sims) G. Don = 3-1
    sphalera Leeuwenb., see 10, type
Fimbrolina
    incarnata (Aubl.) Raf. = 19-1
Fritschiantha
    purpurascens (Rusby) Kuntze = 11-2

Gesneria
   aurantiaca Hanst. = 19-1
   guianensis Benth. = 19-2
   hirsuta Kunth = 12-1; see 12, type
   schomburgkiana Kunth & Bouché = 19-2
Gloxinia
   maculata L'Her., see 11, type
   suaveolens Decne. = 11-1
   sylvatica (Kunth) Wiehler, see 11, type
   trichantha Miq. = 11-1
Hypocyrta
   sect. *Codonanthe* Mart. = 4
   ciliosa Mart. = 17-4
   crassifolia Focke = 4-2; see 4-2, note
   gracilis Mart., see 4, type
Isoloma
   hirsutum (Kunth) Regel = 12-1
Kohleria
   hirsuta (Kunth) Regel var. longipes (Benth.) L.P. Kvist & L.E. Skog,
      see 12-1, note
   tubiflora (Cav.) Hanst., see 12, note
Lophalix
   coccinea (Aubl.) Raf. = 9-2
Macrochlamys
   patrisii (DC.) Decne. = 9-2
Martynia
   perennis L. = 11-1; see 11, type
Napeanthus
   brasiliensis Gardner, see 15, type
   primulifolius (Raddi) Sandwith, see 15, type
Nautilocalyx
   bracteatus (Planch.) Sprague, see 16, type
   cordatus (Sprague) L.E. Skog, see 16, type
   hastatus Linden ex Hanst., see 16, type
   hirsutus (Sprague) Sprague, see 16-4
   lacteus Sandwith = 16-9
   melittifolius (L.) Wiehler, see 16, note
   villosus (Kunth & Bouché) Sprague, see 16, note
Nematanthus
   calcaratus Miq. = 4-1
   savannarum (C.V. Morton) J.L. Clark, see 1-2, note
Ortholoma (Benth.) Hanst. = 6
   acuminatum (Benth.) Hanst., see 6, type
   calotrichum (Donn. Sm.) Wiehler = 6-1

Paradrymonia
  ciliosa (Mart.) Wiehler, see 17, type
  glabra (Benth.) Hanst. = 17-4; see 17, type
Pentadenia (Planch.) Hanst. = 6
  aurantiaca (Decne. ex Planch.) Hanst., see 6, type
Rechsteineria
  aurantiaca (Hanst.) Kuntze = 19-1
  crenata Fritsch = 19-2
  faucidens Hoehne var. *parvifolia* Hoehne = 19-1
  incarnata (Aubl.) Leeuwenb. = 19-1
  schomburgkiana (Kunth & Bouché) Kuntze = 19-2
Rhoogeton
  leeuwenbergianus C.V. Morton = 18-2
Saintpaulia spp., see family, note; see 19, note
Salisia
  suaveolens (Decne.) Regel = 11-1
Seemannia Regel = 11
  purpurascens Rusby = 11-2
  ternifolia Regel, see 11, type
Sinningia
  helleri Nees, see 19, type
  speciosa (Lodd.) Hiern, see family, note; see 11, note; see 19, note
Skiophila
  pulchella (Donn ex Sims) Hanst. = 3-1
Trichantha Hook. = 6
  calotricha (Donn. Sm.) Wiehler = 6-1
  minor Hook., see 6, type
Trichanthera
  gigantea (Humb. & Bonpl.) Nees, see 2, note
Tussacia
  pulchella (Donn ex Sims) Benth. = 3-1
  rupestris Benth. = 3-2
  villosa Benth. = 3-1
Tylosperma Leeuwenb. = 20
  cuneatum (Gleason) Leeuwenb. = 20-1

INDEX TO VERNACULAR NAMES

Alphabetic list of families of series A occurring in the Guianas

Defined as in Cronquist, 1981, and numbered in his sequence, with alternative names. Those published, with chronological fascicle number and year.

Abolbodaceae		
(see Xyridaceae	182)	15. 1994
Acanthaceae	156	23. 2006
(incl. Thunbergiaceae)		
(excl. Mendonciaceae	159)	
Achatocarpaceae	028	22. 2003
Agavaceae	202	
Aizoaceae	030	22. 2003
(excl. Molluginaceae	036)	22. 2003
Alismataceae	168	
Amaranthaceae	033	22. 2003
Amaryllidaceae		
(see Liliaceae	199)	
Anacardiaceae	129	19. 1997
Anisophylleaceae	082	
Annonaceae	002	
Apiaceae	137	
Apocynaceae	140	
Aquifoliaceae	111	
Araceae	178	
Araliaceae	136	
Arecaceae	175	
Aristolochiaceae	010	20. 1998
Asclepiadaceae	141	
Asteraceae	166	
Avicenniaceae		
(see Verbenaceae	148)	4. 1988
Balanophoraceae	107	14. 1993
Basellaceae	035	22. 2003
Bataceae	070	
Begoniaceae	065	
Berberidaceae	016	
Bignoniaceae	158	
Bixaceae	059	
(incl. Cochlospermaceae)		
Bombacaceae	051	
Bonnetiaceae		
(see Theaceae	043)	
Boraginaceae	147	
Brassicaceae	068	
Bromeliaceae	189	p.p. 3. 1987
Burmanniaceae	206	6. 1989
Burseraceae	128	
Butomaceae		
(see Limnocharitaceae	167)	
Byttneriaceae		
(see Sterculiaceae	050)	
Cabombaceae	013	
Cactaceae	031	18. 1997

Caesalpiniaceae	088	p.p. 7. 1989
Callitrichaceae	150	
Campanulaceae	162	
(incl. Lobeliaceae)		
Cannaceae	195	1. 1985
Canellaceae	004	
Capparaceae	067	
Caprifoliaceae	164	
Caricaceae	063	
Caryocaraceae	042	
Caryophyllaceae	037	22. 2003
Casuarinaceae	026	11. 1992
Cecropiaceae	022	11. 1992
Celastraceae	109	
Ceratophyllaceae	014	
Chenopodiaceae	032	22. 2003
Chloranthaceae	008	24. 2007
Chrysobalanaceae	085	2. 1986
Clethraceae	072	
Clusiaceae	047	
(incl. Hypericaceae)		
Cochlospermaceae		
(see Bixaceae	059)	
Combretaceae	100	
Commelinaceae	180	
Compositae		
(= Asteraceae	166)	
Connaraceae	081	
Convolvulaceae	143	
(excl. Cuscutaceae	144)	
Costaceae	194	1. 1985
Crassulaceae	083	
Cruciferae		
(= Brassicaceae	068)	
Cucurbitaceae	064	
Cunoniaceae	081a	
Cuscutaceae	144	
Cycadaceae	208	9. 1991
Cyclanthaceae	176	
Cyperaceae	186	
Cyrillaceae	071	
Dichapetalaceae	113	
Dilleniaceae	040	
Dioscoreaceae	205	
Dipterocarpaceae	041a	17. 1995
Droseraceae	055	22. 2003
Ebenaceae	075	
Elaeocarpaceae	048	
Elatinaceae	046	

Eremolepidaceae	105a	25. 2007
Ericaceae	073	
Eriocaulaceae	184	
Erythroxylaceae	118	
Euphorbiaceae	115	
Euphroniaceae	123a	21. 1998
Fabaceae	089	
Flacourtiaceae	056	
(excl. Lacistemaceae	057)	
(excl. Peridiscaceae	058)	
Gentianaceae	139	
Gesneriaceae	155	26. 2008
Gnetaceae	209	9. 1991
Gramineae		
(= Poaceae	187)	8. 1990
Gunneraceae	093	
Guttiferae		
(= Clusiaceae	047)	
Haemodoraceae	198	15. 1994
Haloragaceae	092	
Heliconiaceae	191	1. 1985
Henriquesiaceae		
(see Rubiaceae	163)	
Hernandiaceae	007	24. 2007
Hippocrateaceae	110	16. 1994
Humiriaceae	119	
Hydrocharitaceae	169	
Hydrophyllaceae	146	
Icacinaceae	112	16. 1994
Hypericaceae		
(see Clusiaceae	047)	
Iridaceae	200	
Ixonanthaceae	120	
Juglandaceae	024	
Juncaginaceae	170	
Krameriaceae	126	21. 1998
Labiatae		
(= Lamiaceae	149)	
Lacistemaceae	057	
Lamiaceae	149	
Lauraceae	006	
Lecythidaceae	053	12. 1993
Leguminosae		
(= Mimosaceae	087)	
+ Caesalpiniaceae	088)	p.p. 7. 1989
+ Fabaceae	089)	
Lemnaceae	179	
Lentibulariaceae	160	
Lepidobotryaceae	134a	
Liliaceae	199	
(incl. Amaryllidaceae)		
(excl. Agavaceae	202)	
(excl. Smilacaceae	204)	
Limnocharitaceae	167	
(incl. Butomaceae)		
Linaceae	121	
Lissocarpaceae	077	

Loasaceae	066	
Lobeliaceae		
(see Campanulaceae	162)	
Loganiaceae	138	
Loranthaceae	105b	25. 2007
Lythraceae	094	
Malpighiaceae	122	
Malvaceae	052	
Marantaceae	196	
Marcgraviaceae	044	
Martyniaceae		
Mayacaceae	183	
Melastomataceae	099	13. 1993
Meliaceae	131	
Mendonciaceae	159	23. 2006
Menispermaceae	017	
Menyanthaceae	145	
Mimosaceae	087	
Molluginaceae	036	22. 2003
Monimiaceae	005	
Moraceae	021	11. 1992
Moringaceae	069	
Musaceae	192	1. 1985
(excl. Strelitziaceae	190)	
(excl. Heliconiaceae	191)	
Myoporaceae	154	
Myricaceae	025	
Myristicaceae	003	
Myrsinaceae	080	
Myrtaceae	096	
Najadaceae	173	
Nelumbonaceae	011	
Nyctaginaceae	029	22. 2003
Nymphaeaceae	012	
(excl. Nelumbonaceae	010)	
(excl. Cabombaceae	013)	
Ochnaceae	041	
Olacaceae	102	14. 1993
Oleaceae	152	
Onagraceae	098	10. 1991
Opiliaceae	103	14. 1993
Orchidaceae	207	
Oxalidaceae	134	
Palmae		
(= Arecaceae	175)	
Pandanaceae	177	
Papaveraceae	019	
Papilionaceae		
(= Fabaceae	089)	
Passifloraceae	062	
Pedaliaceae	157	
(incl. Martyniaceae)		
Peridiscaceae	058	
Phytolaccaceae	027	22. 2003
Pinaceae	210	9. 1991
Piperaceae	009	24. 2007
Plantaginaceae	151